D0498203

Fire

FIRE

John W. Lyons

An imprint of Scientific American Books, Inc.
New York

Library of Congress Cataloging in Publication Data

Lyons, John W., 1930–
Fire.

(Scientific American library)
Includes index.
1. Fire. I. Title. II. Series.
TP265.L874 1985 621.402′3 85-2185
ISBN 0-7167-5010-4

Printed in the United States of America

Book design by Malcolm Grear Designers

Scientific American Library is published by Scientific American Books, Inc., a subsidiary of Scientific American, Inc.

Distributed by W. H. Freeman and Company, 41 Madison Avenue, New York, New York 10010

This is number 14 of a series. Earlier volumes in the series are unnumbered.

1 2 3 4 5 6 7 8 9 KP 3 2 1 0 8 9 8 7 6 5

To the memory of Louis M. Lyons, 1897–1982

Contents

Preface

The ideas in this book have been long developing. Many years ago, after writing a monograph on the chemistry and uses of fire retardants, as well as managing an extensive program on fire research, I thought to write a book summarizing the entire field of fire research. I found the subject too broad to cover in a single book at the technical level. Then the editors of the Scientific American Library asked me to try a less detailed but even broader discussion aimed at a general audience.

Such an assignment proved difficult, given the complicated nature of fire. I was challenged by the duality of the subject itself: fire is both the servant of humanity and its destructive demon. I tried to present both aspects, comprehensively, so about half the book now treats intentional fires and about half fires out of control. Another major challenge was to describe the appropriate physics, chemistry, and engineering concepts without much mathematics. This was a difficult task for one trained in the shorthand of scientific notation; I hope I have succeeded. My colleagues will appreciate, for example, the problem of describing Fourier's contributions to physics without going into extensive mathematics.

For the ancients, fire held a special place. To them the ever-changing motion and colors so evident in flames made fire a symbol of life and even of the presence of gods. Candles still play an important symbolic role in most religions. The ancients appointed certain gods to guard the hearth and prevent the fire from going out—in the days before friction matches, fire could be very difficult to produce. Other gods tended volcanos and other evil fires. The Greeks devised elaborate constructs of the earth and universe based on the "four elements": earth, water, air, and fire. One system had the earth surrounded by concentric shells of air, water clouds, and fire. Rents in the water cloud revealed the stars as visible portions of the fire shell. Some of this Greek thought persisted through the days of the alchemists and astrologers into the 1770's, when Lavoisier refuted the phlogiston theory.

Scientists and natural philosophers have always been intrigued by fire. In this book I shall discuss the efforts of a few of these men—Lavoisier, Rumford, Faraday, Watt, and Franklin—to show how substantial theories of science and engineering have grown out of an interest in fire and combustion. Our own attitude toward fire reflects this old fascination as we dream in front of a log fire or gaze into the flickering flame of a candle.

I am indebted to many people. My family has put up with the litter of this manuscript for too long. Mary Chandler typed one whole draft; my daughter Peggy, another. Karen McDermott at Scientific American Books went through the copy-editing stages with me and was my principal collaborator in the production of the book. David Barkan was photo researcher, and Gabor Kiss produced splendid drawings. And finally, Gerard Piel went through the entire manuscript, making very significant contributions to style, clarity, and content, recasting and rewriting a good deal of the material. My many colleagues at the National Bureau of Standards provided references, photographs, and helpful discussions. To all these and to others involved at earlier stages goes my deepest appreciation.

John W. Lyons
Mt. Airy, Maryland
March 1985

Fire

1 *Fire*

Fire provides the material well-being of the people in the industrial countries of the world. Heat from the burning of fuel converted into electrical and mechanical energy does practically all the work of these economies. The fuel burned by the United States in 1980 produced 335 million British thermal units per capita. That was enough heat to bring to a boil more than 250,000 gallons of water for each person. The American people found better uses than that, of course, for all this heat. From the water that they did boil—in central power stations—they extracted more than 10,000 kilowatt-hours of electrical energy apiece. If they transformed the other two-thirds of the available Btu's at just one-fourth the efficiency of those central power stations, they enjoyed another 5,000 kilowatt-hours of goods and services. If one reckons a man-year at 150 kilowatt-hours, in the U.S. economy in 1980 the fires that burned the equivalent of 8.5 billion tons of coal, in petroleum, natural gas, and coal, served each American with the equivalent of 100 slaves.

Yet the American people have a fire problem. Along with their Canadian neighbors they lose property and life to fire at twice the rate of people in comparable circumstances in other industrial countries. Strange to say, this violent misfortune overtakes them most often when they are away from the places where fire does their work. Less than 5 percent of fatal fires happen in nonresidential buildings; about 10 percent must be charged to the much larger highway-accident toll. Two-thirds of all fatal fires occur in the freestanding single-family or two-family house.

Along with the 5,000 Americans who perish in these fires, another 300,000 suffer sufficiently serious burns to require medical treatment. Of these, 30,000 are hospitalized for prolonged surgical and medical care. Physicians testify that a major burn is the most serious insult the human body can experience. Progress in treating the wound(s) can be overwhelmed by systemic breakdown involving primary salt and fluid balances, the kidneys, the liver, and the adrenal glands; shock, pneumonia, and uremic poisoning are constant antagonists. The pain the victims must endure is so harrowing that medical staff have difficulty working for long on burn wards.

Places of Fire Fatalities in the United States (1982)

Residential		4,940
One- and two-family dwellings	3,960	
Apartments	860	
Hotels and motels	75	
Other residential	45	
Nonresidential structures		260
Highway vehicles		575
Other vehicles		120
All other		125
Total		6,020

Source: Estimates provided by the National Fire Protection Association.

Fire specialists have no ready answer why the North American experience with fire is so much worse than that of people elsewhere in the world. Some argue that culture and attitude are important. Here the contrast is most often drawn with the Japanese. The fact is that Japanese losses to urban conflagration attending great catastrophes from other causes, even in this century, greatly exceed those of other peoples. Nonetheless, their annual losses to fire

Young Japanese join in a fire-suppression exercise done in connection with a large-scale disaster management drill conducted periodically in Japan in anticipation of a major earthquake.

run at half the American experience. The great disasters, such as the Tokyo earthquake and conflagration of 1923, have taught the Japanese to treat fire with discipline and respect. Until very recently, a fire on one's own premises was prosecuted as a crime; it remains a serious social offense: the endangerment of neighbors.

It is argued, further, that a high incidence of fire correlates positively with gross national product, high energy consumption, and other measures of economic well-being. Affluence increases, surely, the amount of combustibles in the home, including carpeting, upholstered furniture, and wood-paneled walls. Europeans have been overtaking Americans in standard of living, however, without a corresponding increase in the cost they pay to fire, and the Japanese have made substantial recent gains in per capita income while holding steady their annual loss of life and property to fire.

Many people in this country, a larger number a few years ago than now, credit the difference to the willingness of people to have their government intervene in their lives. The Japanese are cited again as the case in point. In their equable climate and only recent accession to affluence, the Japanese have depended on kerosene space heaters; earthquakes have started thousands of simultaneous fires from these heaters in Japanese cities. Some years ago heaters were devised with a shutoff triggered by rapid sideward movement or overturning. Citizens in Tokyo were given a short notice period to replace their old heaters with new ones or face a fine. In European countries as well, social controls on property and behavior are more rigorous. It may be no accident that the North Americans with their frontier spirit of rugged individualism have the worst fire record.

In historical perspective the fire problem of the Americans can be regarded as the residual of a much larger and more universally besetting problem of urban civilization. In earlier times it was not one house afire, but a whole city. Conflagration has marked the history of every European city, the Great Fire of London in 1666 being most often recalled by Americans. Yet the London fire brigade did not receive city tax support until the 1860's; until then, fire protection of the city had been financed by property owners and insurance underwriters. The technology for suppressing city fires made such progress that those in the 20th century have happened only in consequence of wider natural or manmade calamities.

Much of the progress in preventing city fires came from the technologies that provide city-wide high-pressure water mains and fire-resistant buildings of stone, reinforced concrete, and steel. Beginning in about 1900, attention shifted to the development of designs and test methods to achieve specified levels of fire resistance within individual buildings. Incorporated in the building codes of local communities and states in this country over the past 50 years, such standards have been built into all large buildings. Their columns and beams meet specified levels of fire resistance, fire-rated walls separate

Fire-Caused Fatalities in the United States

Year	Deaths	Deaths per 100,000 population
1950	6,405	4.2
1952	6,922	4.4
1954	6,003	3.7
1956	6,405	3.8
1958	7,291	4.2
1960	7,645	4.2
1962	7,534	4.0
1964	7,379	3.8
1966	8,084	4.1
1968	7,335	3.7
1970	6,718	3.3
1972	6,714	3.2
1974	6,236	2.9
1976	7,480	3.5
1978	7,440	3.4
1979	6,245	2.8
1980	5,765	2.6
1981	5,860	2.6
1982	5,325	2.3
1983	5,195	2.2

Sources: Data for 1950–1974 from U.S. Government, National Center for Health Statistics. Data for 1976–1983 from National Fire Protection Association (with vehicular fatalities subtracted).

Fire-Caused Fatalities in Various Nations

Nation	Deaths per 100,000 population 1974*	1976–1978 (av.)†	Latest report†
Canada	3.6	3.2	2.9
United States	2.9	2.9	2.8
Sweden	1.6	1.5	1.6
Japan	1.5	—	1.5
United Kingdom	1.5	1.5	1.5
France	1.5	1.5	1.5
Australia	1.5	—	0.8
West Germany	0.9	0.9	0.9
Switzerland	0.7	0.6	0.7

*Data from W. Berl and B. Halpin, Johns Hopkins Applied Physics Laboratory, Laurel, Maryland.
†Published and unpublished data from Philip Schaenman, Tri Data, Arlington, Virginia.

major areas, fire doors isolate stairwells, and so on. Because of these efforts, an entire building is now rarely lost to a fire. The effect in the United States has been to reduce, on a per capita basis, property loss since 1900 fourfold and loss of life by half.

Emphasis in the development of fire-prevention technology has been turning, therefore, to reducing the combustibility of the contents of buildings. Progress in this line can contribute substantially to amelioration of the residential fire problem, if people can be persuaded to weigh this factor in furnishing and decorating their homes. We cannot count on unaided behavioral changes, however, to solve the problem. A recent study by the Federal government of Census Bureau data shows there are approximately 5 million fires in U.S. households each year, about 1 million of them serious enough to call the fire department. Among the 70 million households, that is one chance in 14. To achieve a reduction in U.S. fire losses by half, a goal set by the U.S. Congress in the Federal Fire Prevention and Control Act of 1974, we shall have to bring the level of concern for fire safety in the home up close to that long established for public safety in large buildings.

Now that protection of life in the home has become the central concern, fire-prevention technology must address the earliest states of the ignition and growth of the unwanted fire. As will be shown later in this book, most deaths

Ten of the Greatest Conflagrations in Recent History*

Year	Deaths	Conflagration
1923	200,000†	Tokyo earthquake and fire
1657	2,574	Edo (Tokyo)
1772	1,551	Edo (Tokyo)
1724	1,231	Osaka
1906	1,188	San Francisco
1923	990	Yokohama
1871	776	Chicago
1864	647	Kyoto
1936	428	Hakodate
1666	17	London

*Conflagrations created by wartime bombings have not been included in this table. Firestorms created by Allied bombings killed about 21,000 in Hamburg, 135,000 in Dresden, and 84,000 in Tokyo; a nuclear attack today could turn every major city into a firestorm.
†Includes all deaths (only part of them were from fires); the San Francisco and Yokohama fires were caused by earthquakes, but only fire-related deaths are included here. (All of the figures except those for the 1923 Tokyo disaster are from the U.S.-Japan National Resources Panel on Fire Research, 1976.)

result at the time of the fire from inhalation of the combustion products of the fire. We must learn how these gases are produced and distributed from a fire in a typical dwelling, how to detect such gases and quickly alert occupants, and how to provide fast, sure, and simple means of escape. Phenomena to be understood include ignition, smoldering combustion, the early stages of flaming combustion and its spread, and gas dynamics in rooms and assemblages of rooms and corridors.

Modern fire technology, therefore, engages fundamental studies of the physics and chemistry of flames, of heat transfer, of fluid mechanics, and of the chemistry of materials. For consideration of what we have learned and have yet to learn about these matters, this book begins with a closer look at fire in its friendly role in our lives—in the candle, in the fireplace, in furnaces and combustion chambers—before it reviews our uneven contest with the primal violence of fire out of control.

Heat radiation registered on infrared film in a camera aboard a high-altitude photographic satellite plainly maps the concentration of the U.S. population in the eastern half of the country. Heat from fire in electric-utility furnaces and combustion chambers supplies each American with an average of 100 mechanical slaves, measured in 150 kilowatt-hour man-year equivalents.

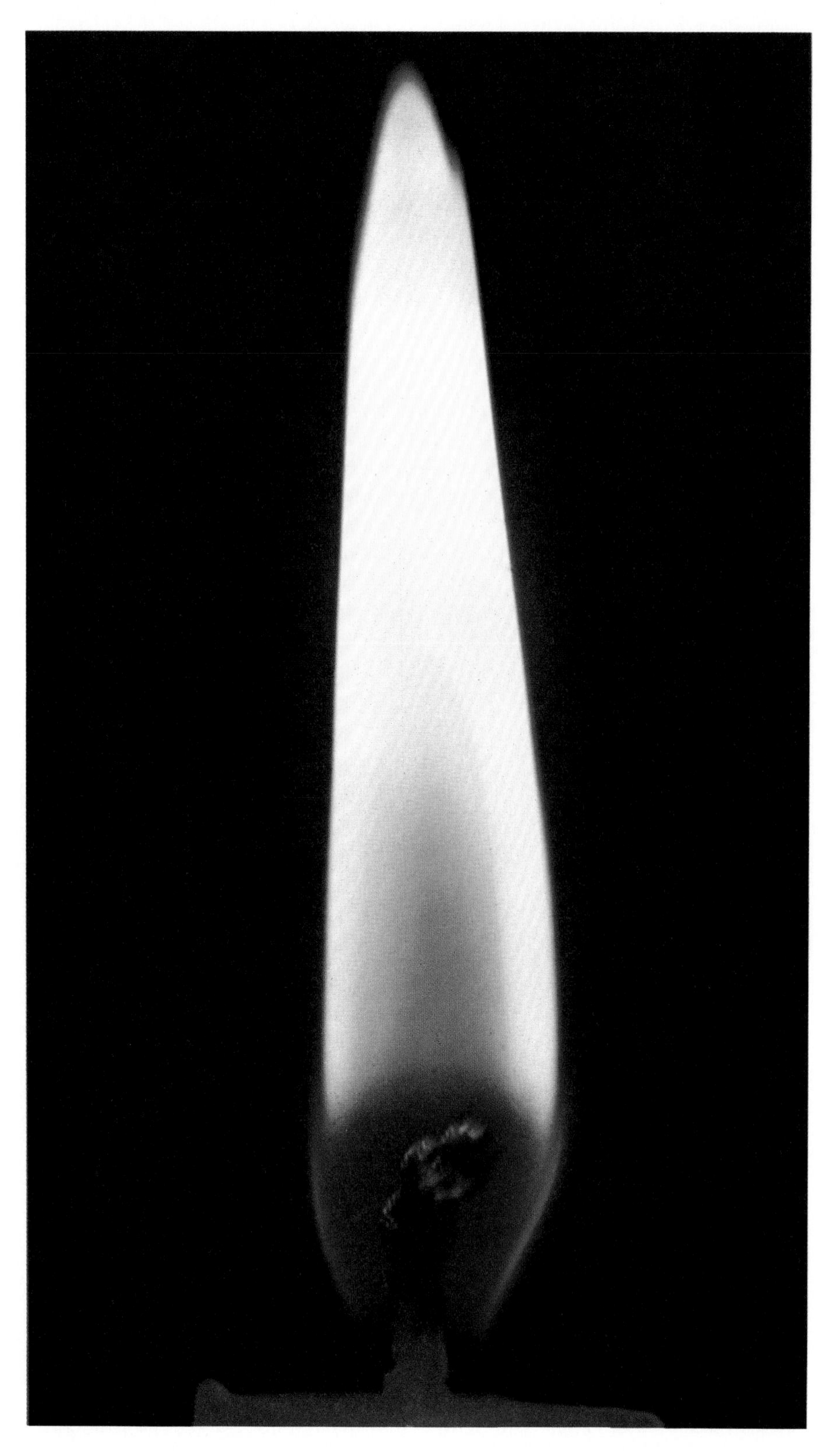

The candle flame invites inspection. Its structure reflects the physics and chemistry of the combustion reactions by which it makes its light.

2 *The Candle*

The candle, these days, has a purely decorative role in indoor lighting. Its displacement by the electric light bulb began with an experiment performed in 1831 by Michael Faraday. In his laboratory at the Royal Institution in London, Faraday moved a wire between the poles of a magnet and thereby induced the flow of an electric current. He was concerned not with the improvement of indoor lighting but with the nature of the forces of electricity and magnetism. Faraday's experiment established the intimate relation of the two forces and opened new realms of nature to physics. The fruits of that same experiment, the induction of electricity running continuously in dynamos all around the world, have transformed human existence.

The candle has yet another role today: as an object of fascination and instruction in physics. This role again it owes to Michael Faraday. It was and continues to be the custom of the Royal Institution to invite the interested public, especially students, to hear an annual series of Christmas lectures and to witness the sometimes spectacular accompanying demonstrations. Faraday gave his lectures titled "The Chemical History of a Candle" in 1848 and, in an encore, in 1860. Transcribed very much as he gave them, with his concurrent description and explanation of his demonstrations, these lectures continue in print, a classic in the popular literature of science. In Faraday's words:

> There is no more open door by which you can enter into the study of natural philosophy than by considering the physical phenomena of a candle. There is not a law under which any part of this universe is governed which does not come into play, and is not touched upon, in these phenomena.

For the more limited practical objective of this book, the understanding, management, and control of fire, let us now consider the physical phenomena of Faraday's candle. The physical and chemical transactions responsible for its remarkable behavior will serve us in good stead as we encounter fire in its more formidable manifestations further on.

From time to time it will be helpful if the reader puts down this book and observes a candle flame for a while. Fix a well-shaped candle of good quality in a steady holder and set up a book of matches.

Now, light the candle.

The first thing to remark is the steady, balanced state of the flame. It reaches that state the moment it catches on, the moment the wax melts in a pool at the top of the candle. If the room air is reasonably still, the flame does little flickering. It neither dies down nor grows in size. It remains constant in height and shape and burns a fixed distance above the pool of melted wax. In a well-made candle, even the relative length and the curl of the somehow self-trimming wick, visible inside the flame, remain constant. Everything is in balance: wax is consumed, light is emitted, and heat is produced in a condition termed a steady state. It is this serene balance that makes the candle seem so remarkable; indeed, gives it a hypnotic effect. About that effect there is, undoubtedly, much to learn. We have set ourselves the task of unraveling, however, the physical and chemical phenomena that underlie it.

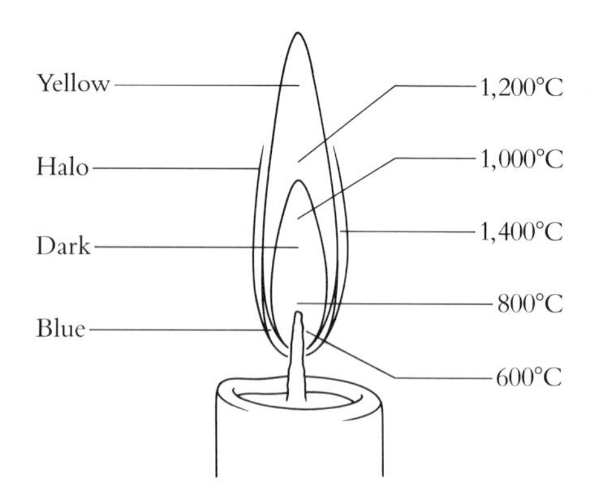

Zones of color and of temperature in the candle flame give clues to what is going on in it. The hottest temperatures are in the blue reaction zone near the bottom and upward on the surface of the flame. The coolest is the dark zone into which wax is vaporizing from the wick. Most of the light comes from incandescent soot particles as they burn away in the yellow tongue.

Closer observation suggests new questions that can be answered in part by a few simple experiments such as Faraday performed for his audience. Thus, when you light the candle, notice that the pool of melted wax forms immediately. The flame cannot sustain itself without this molten pool. The flame seems to touch the wick but not the surface of the pool. If you snip off the wick, you cannot light the top of the candle; the wax alone will not sustain a flame.

Now look at the steady-state flame, and notice that it has a structure. The brightest part of the flame, the yellow luminous region that starts at the shoulder of the curl of the wick and tapers off at the top, catches the eye first. This region encloses a dark cone that reaches up into it, starting from well below the curl of the wick. At the bottom of the flame around the wick, but not quite touching it, you can see a blue region, also of lesser brightness, that seems to cup and envelop the base of the dark and the yellow regions, reaching thinly up into the yellow until the eye loses it in the light.

If, following Faraday, you put a cold spoon into the bright yellow part of the flame, say two-thirds of the way up, you will find that it is instantly covered with black soot. Then take the spoon, with some cold water in it but dry on the bottom, and hold it about two inches above the flame. You will discover that water has condensed on the bottom. (The effect is fleeting because the heat of the flame quickly warms the spoon above the dew point.) So, just above the flame there is no blackening of the spoon. The soot particles have vanished.

Faraday performed some other experiments that required a little more equipment than is usually available in a drawing room or kitchen. He placed the end of a thin, bent glass tube in the dark region of the flame and showed, by touching a match to the other end, that vaporized but not yet burning wax rushes through this region. When he placed the end of the tube in the bright region, a curl of sooty smoke issued from the other end. But smoke, too, may be a fuel. Faraday showed this by quickly placing a match in the smoke of a snuffed-out candle. When this is done quickly enough, the smoke ignites and flashes back to the wick. You can do this with one or two tries.

With the help of such experiments, it quickly becomes clear that structure in the candle flame, as so often elsewhere, is related to function, to what is going on in the flame. We can now visualize the flows of mass and of heat, matter and energy, through the structure. The wax in the solid state warms, melts, moves up the wick, vaporizes from the heat of the flame, and burns. As the distribution of temperature in the flame indicates (*see the figure on page 8*), it is in the blue zone that the principal burning proceeds. There the wax vapor issuing from the wick and diffusing across the dark zone makes contact with oxygen diffusing in from the surrounding air. The resulting combustion reactions generate most of the heat in the flame. The products of these reactions flow up into the yellow, luminous tongue of the flame, where burning proceeds to completion. Most of the light comes from this zone in the flame; it comes from the sooty carbon particles heated to incandescence and entering into combustion with oxygen from the air. Water vapor and carbon dioxide, the final combustion products, issue in an invisible plume from the tip.

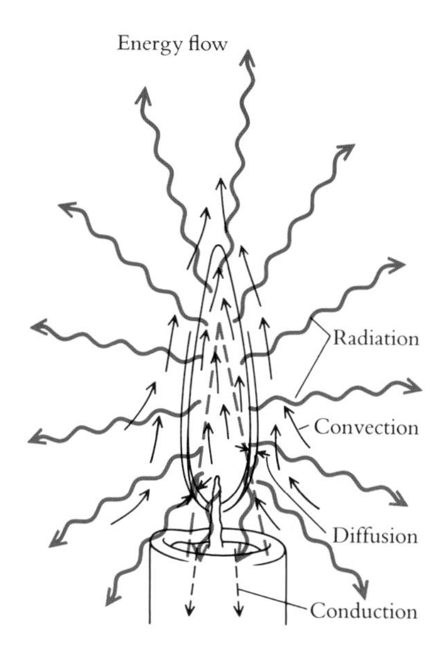

Radiation (wavy arrows) carries off about 25 percent of the total energy of combustion equally in all directions from the flame. The roughly 4 percent of the radiation that feeds back to the candle melts the wax to fuel the flame. The melted wax moves upward through the wick by capillary action. Vaporized there by intense radiation from the surrounding reaction zone, the wax molecules go into cracking and combustion reactions that lead ultimately to oxidation of their constituent carbon atoms to carbon dioxide and water.

The heat flows are more complex. They may be described as mechanical; that is, expressed in motion of particles, and electromagnetic; that is, by radiation. The radiant heat issues equally in all directions, or isotropically. Downward it melts the wax in the pool at the base of the wick. Converging inward from the reaction zone around the dark region at the bottom of the flame, it vaporizes the wax in the wick and heats the vapor to combustion temperature.

The flow of the heat that is absorbed by particles and converted into the energy of motion goes mainly upward. But some of the heat is carried downward by conduction. This is the transfer of energy by collision from more excited to less excited particles. In the candle such transfer proceeds throughout the flame but most significantly for the maintenance of the flame from the hot pool of liquid wax downward into the solid wax.

Some of the heat moves by thermal diffusion, the actual movement of particles from hotter regions (where velocities are faster) to cooler regions (where velocities are slower). Diffusion is the result of the ability of many freely moving particles in a system to mix thoroughly in a fairly short time. Thus if one uncorks a flask containing a little mercaptan (the substance that gives skunks their odor) in a room and then opens the door to the next room, the odor will soon be evident in the second room. Before long the intensity of the smell will be the same everywhere. This mixing is driven by the difference in the concentrations of the gas molecules of the mercaptan and the molecules of air and is called chemical diffusion. In contrast, thermal diffusion is driven by differences in temperature and so takes place even when the molecules or particles are all the same chemical type. Chemical diffusion is very important in flames, as we shall see.

Whichever way they happen to be moving by conduction and diffusion, the atoms and molecules in the flame are all entrained in the upward motion of convection. The hot, expanding gases are lighter than the surrounding air and so ascend buoyantly.

The downward and inward flow of heat by radiation and some conduction is crucial to the maintenance of the flame. The feedback of a small portion of the heat generated by the flame keeps the solid wax melting into the pool at the base of the flame at just the right rate. Similarly, the radiation from the blue region vaporizes the melted fuel in the wick to supply fuel for the primary combustion reactions there and in the light-generating yellow zone above. The secret of the balance of effect in the candle is that the amount of heat reaching the candle top and wick is just enough to melt and vaporize the next increment of fuel. If we could suddenly increase the heat of combustion, the effect would be an increase in the melting and vaporization rate, producing more fuel and a hotter and larger flame. The effect would be throttled by two factors, the rate of movement of liquid wax up the wick and the diffusion of fuel outward, and of oxygen inward, to the reaction zone. Thus the flame would stabilize at a new and larger shape and size; it is a self-adjusting system.

The experiments and observations that have helped us to see this much in the flame give rise to many more questions than they answer. That is how science makes progress. Without waiting for answers, artisans learned over the years to make better candles. By empirical, try and fail and try again methods, they learned that candles are best made from a range of waxy materials that melt at moderate temperatures. They found that the wick must have an open structure so that the melted wax can move through it, by capillary action, into the flame. Further experimenting showed the relation of wick diameter and effective wick surface to flame size and what salts to add to wick materials to cause the curling you observe and to produce clean wick trimming by the flame. About waxes and the optimum diameters of candles made from different waxes candlemakers have accumulated, of course, equivalent lore. Making candles that consume themselves in a clean, steady-state flame is a fine art.

The Physics of the Flame

We must now put the lore of the candle aside to consider the underlying physics of fire. We shall see that these principles are well illustrated by its flame.

Until little more than 200 years ago heat was thought to be a substance.

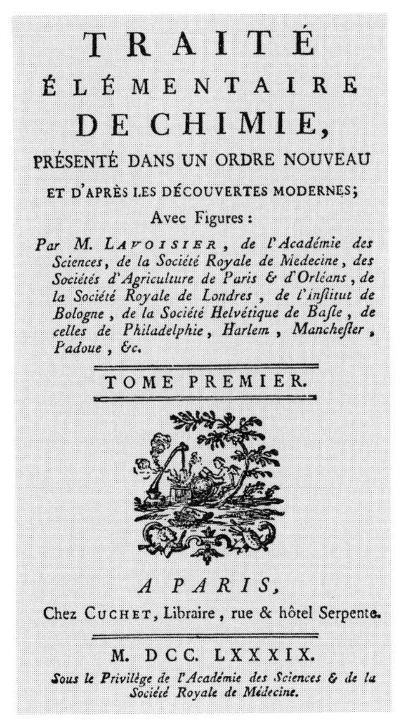
TRAITÉ
ÉLÉMENTAIRE
DE CHIMIE,
PRÉSENTÉ DANS UN ORDRE NOUVEAU
ET D'APRÈS LES DÉCOUVERTES MODERNES;
Avec Figures:
Par M. LAVOISIER, *de l'Académie des Sciences, de la Société Royale de Medecine, des Sociétés d'Agriculture de Paris & d'Orléans, de la Société Royale de Londres, de l'inſtitut de Bologne, de la Société Helvétique de Baſle, de celles de Philadelphie, Harlem, Mancheſter, Padoue, &c.*

TOME PREMIER.

A PARIS,
Chez CUCHET, Libraire, rue & hôtel Serpente.

M. DCC. LXXXIX.
Sous le Privilège de l'Académie des Sciences & de la Société Royale de Médecine.

Title page of Lavoisier's treatise on chemistry. The scientific society at "Philadelphie" cited among his credentials is the American Philosophical Society founded by Benjamin Franklin.

Lavoisier and the Phlogiston Theory

Antoine Laurent Lavoisier was a natural philosopher in the most generous spirit of the 18th-century Enlightenment. Remembered as a chemist, indeed as the founder of modern chemistry, he was a pioneer in scientific agriculture, physiology, and industrial chemistry and a leading figure in finance, economics, education, and government during the years France entered the crisis of its Revolution.

When Lavoisier, at age 29, began his studies of combustion in 1772, chemistry was held in the sway of the phlogiston theory, which postulated that the essence of fire is a substance called phlogiston. The theory of fire as a substance is an ancient idea; fire was one of the four elements of the ancient Greeks. The alchemists propounded a number of ideas about fire and the transformations it brings about, most ill-founded and some fraudulent. The term phlogiston, from this old tradition, became vested with new and sophisticated concepts following some quantitative experimentation on gases by Robert Hooke, Otto von Guericke, and Robert Boyle, all organized in a comprehensive theory by Georg Stahl at the end of the 17th century. The theory explained most combustion transactions. The defects were largely masked by the great difficulties encountered in making accurate weighings and in maintaining tight vacuum seals.

Phlogiston was thought to be the major constituent of good fuels and to be released into the air on burning; a candle or lump of charcoal was thus considered to be nearly pure phlogiston. Air saturated with phlogiston no longer allows a candle to burn in it. Metals contain more phlogiston than their ores or calces (oxides). The winning of a metal from its ore proceeded as follows:

$$\text{metal ore} + \underset{\substack{\text{(from} \\ \text{charcoal)}}}{\text{phlogiston}} \rightarrow \text{metal}$$

On the other hand, the burning of phosphorus was pictured as:

$$\text{phosphorus} - \text{phlogiston} \rightarrow \text{phosphorus calx}$$

By more rigorously exact experiment and measurement, Lavoisier showed that in fact the first transaction yields a loss in weight and the second a gain. The theorists had to posit a negative weight for phlogiston, an untenable position.

In the next step in his inquiry, Lavoisier benefited from a parallel enterprise of Joseph Priestley in England. Priestley had isolated what he called "dephlogisticated air." Adopting Priestley's method of isolating it, Lavoisier demonstrated that it was this substance, which he called "oxygen," that entered into the combustion reaction and added its weight to the calx of phosphorus. The equations could now be rewritten:

metal ore + charcoal = metal + carbon dioxide
phosphorus + oxygen = phosphorus calx

Priestley did not join in the celebration of this discovery, for he remained in the thrall of the phlogiston theory. On the other hand, having disposed of phlogiston, Lavoisier was not satisfied that he had explained the evolution of heat. For this purpose he embraced the caloric theory: caloric, unlike phlogiston an imponderable substance, was said to issue from the combustion reaction.

For his role as manager of the royal franchise that had collected the country's taxes before the Revolution, Lavoisier was guillotined when the Terror took over in 1793. His widow had a brief, unhappy marriage to the social-climbing Benjamin Thompson. Later, as Count Rumford, he showed that Lavoisier's caloric was not a substance but energy transformed from motion into heat.

In the 18th century the argument turned on whether the phlogiston and caloric theories, requiring hypothetical substances of somewhat different behavior, could explain common transformations of materials. Antoine Lavoisier disposed of phlogiston by demonstrating the role of oxygen in combustion, isolating oxygen for the first time and laying the cornerstone of the science of chemistry as he did so. As for the caloric theory, it was Benjamin Thompson, an American soldier of fortune and natural philosopher, who demonstrated in 1797 that heat is a manifestation of motion. As Count Rumford, he commandeered the cannon factory of the Elector of Bavaria as his laboratory for the demonstration (*see the box on pages 14 and 15*).

Rumford's seminal experiment kept some of the greatest scientists of the 19th century engaged in constructing the modern theory of thermodynamics. Sadi Carnot first developed the theory of the macroscopic manifestations of heat in large masses of material. Ludwig Boltzmann, James Clerk Maxwell, and Josiah Willard Gibbs, later in the 19th century, linked the macroscopic phenomena to the microscopic behavior of particles, atoms, and molecules through the powerful conceptual apparatus of statistical mechanics. At the

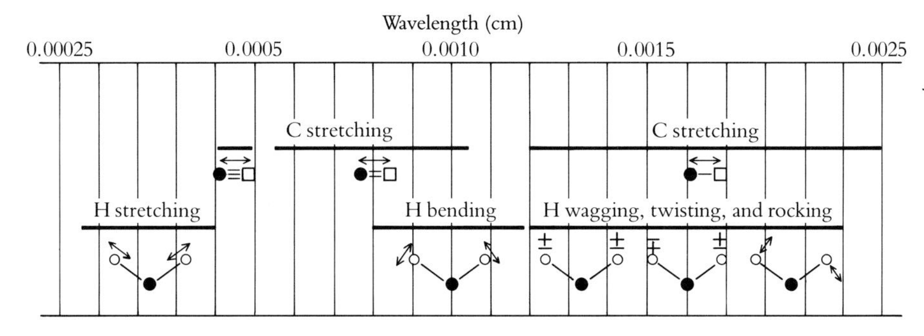

Radiation in the heat or infrared spectrum (see the figure on page 23) originates largely from vibrational motions in intact molecules. A molecule of carbon dioxide here models the kinds of vibrations of its molecular bonds through which it emits radiation at successively shorter wavelengths, from right to left. Carbon atoms are represented by black circles, hydrogen atoms by small open circles, and other atoms by squares. Arrows indicate motion in the plane of the diagram; the + and − signs indicate motion in and out of that plane. The energy carried by the photons emitted increases with decrease in wavelength (see the figure on page 21 and the box on page 22).

turn of this century the quantum mechanics of Max Planck and Albert Einstein successfully filled in many remaining gaps associated with the radiation of light and heat.

A theory of heat requires instruments to measure heat. The thermometers in common use take advantage of the fact that most substances expand as they absorb heat. A temperature scale—Fahrenheit, Celsius, or Kelvin, named for their devisers—arbitrarily marks off even steps between such familiar reference points as the freezing and boiling point of water. For very high temperatures, the temperature dependence of the electrical resistance of platinum is used, and the melting points of gold and silver supply the reference points. The movement of technology into still higher temperature regions keeps thermometry an active field of study globally in measurement laboratories.

At room temperature the wax in a candle is solid. It has, nonetheless, heat content. The heat content of a substance is the amount of internal energy stored in all the possible motions of its constituent atoms and molecules. Such motions persist in every substance down to absolute zero, minus 273.16 degrees Celsius or 0 kelvin, the point at which all motion ceases. The motion encompasses the nucleus and the electrons of the atoms, the vibrations of molecules in many modes, the rotation of molecules and parts of molecules, and, in a solid such as wax, the microscopic movement of molecules, a kind of large-scale vibration.

To increase the temperature of wax takes less energy than to increase the temperature of water. This property, the heat capacity of a substance, is measured on the metric scale as the amount of heat required to raise the temperature of a gram of the substance by one degree Celsius. A material that has a high heat capacity can take up heat through many different modes of motion, whereas a material that has a low heat capacity has few such options. Heat capacity can be better understood by contemplating a large roasting pan and a tall mint julep glass. To fill the pan to a depth of six inches would require a couple of gallons of water; to fill the glass to the same depth would require

Benjamin Thompson, Count Rumford, in 1792.

Benjamin Thompson (Count Rumford)

The ancient Greeks held that heat is a substance, one of the four "elements" with earth, air, and water. This notion began to lose its grip when careful measurements showed that heat is weightless. Benjamin Thompson, from the colonial commonwealth of Massachusetts, delivered the final blow. Thompson is remembered not only as a natural philosopher but also as a social climber, political adventurer, and, at the last, one of the first philanthropic benefactors of science. Compelled to seek his fortune abroad upon his exposure as a Tory sympathizer at the time of the American Revolution, Thompson traveled first to England, then to France, and on to Bavaria. There he ascended in the court of Karl Theodor, Elector of Bavaria, whom he served as counsellor in economic, technological, and political affairs. Named a count of the Holy Roman Empire, Thompson took his title from the early name of the town of Concord, New Hampshire, the home of his first (wealthy) wife. Thus he became Count Rumford.

A highly practical natural philosopher, Thompson turned his studies of convection and radiation to the design of field stoves for the Bavarian army and better fireplaces. Among his inventions were the first drip coffee maker, the sofa bed, and the picture window.

After confirming by careful studies that no weight change is associated with temperature change, he set out to demonstrate that heat is in fact a manifestation of motion. In the cannon factory at the Munich arsenal, Thompson noted the extensive heating of brass cannon barrels while the boring operation was in progress. Then he built a special apparatus, which included a very dull borer, a cannon barrel, and a horse to turn the barrel, to measure the heat generated by the friction. He studied the effect with the apparatus open to air, with air flow prevented, and with the borer and barrel underwater. After two hours and 20 minutes of the barrel turning against the spring-loaded borer at some 32 revolutions per minute, the water actually boiled. (Here he used an exclamation point in his report.) Thus he showed that heat is supplied as long as the horse is turning the cannon barrel.

Reporting this work in 1797, Thompson concluded with one of the most significant statements in the history of science:

> It's hardly necessary to add that anything which any *insulated* body, or system of bodies, can continue to furnish *without limitation,* cannot possibly be a *material substance;* and it appears to me to be extremely difficult, if not quite impossible, to form any distinct idea of anything capable of being excited and communicated in the manner in which heat was excited and communicated in these experiments, except it be MOTION.

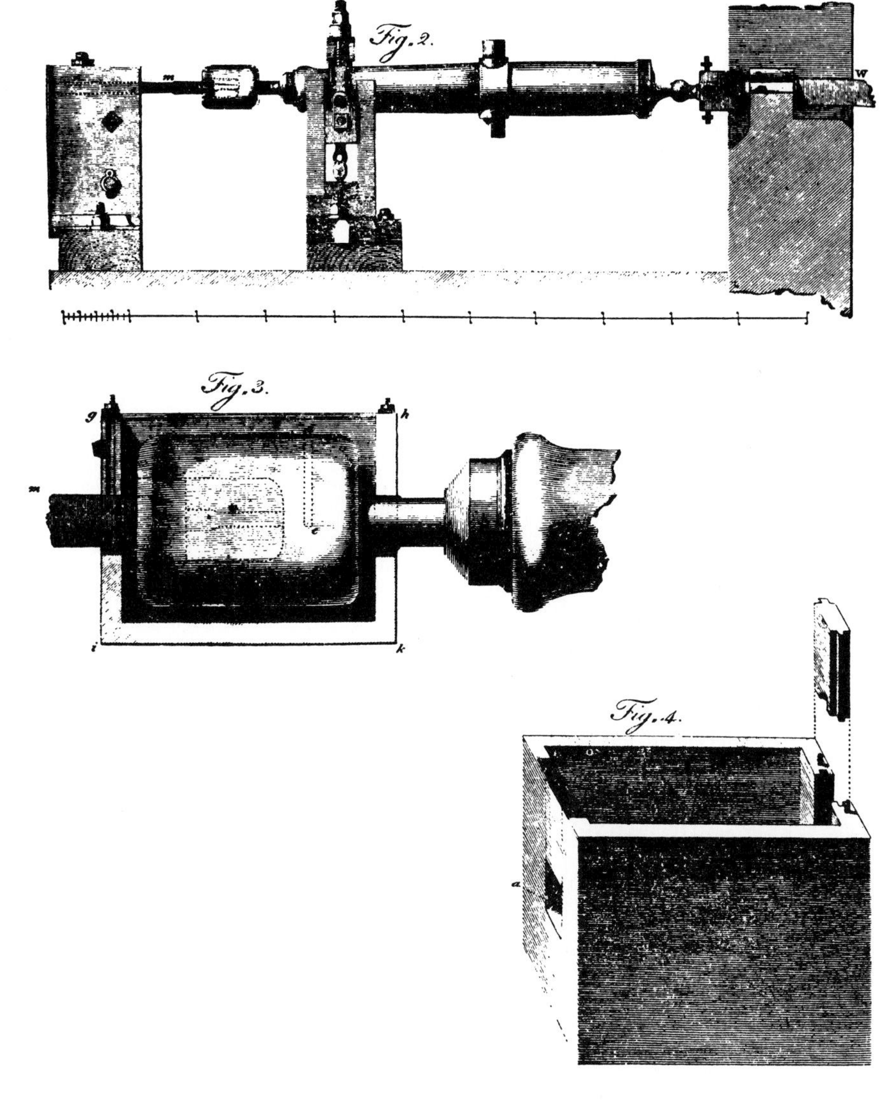

Rumford employed a cannon-boring lathe ("Fig. 2") in the armory of the Elector of Bavaria to disprove the theory of heat as a weightless substance, "caloric," and establish it as "MOTION" (Rumford's emphasis). He set up a dummy workpiece to be turned against a dull tool ("Fig. 3") that would generate friction rather than cut the metal. Immersing the assembly in water in a tank ("Fig. 4"), he brought the water to a boil with the heat of friction.

By doing much the same sort of experiment in the 1840's James Joule established a more precise connection between units of heat and units of work that gives us the conversion factor known as the mechanical equivalent of heat.

Thompson established the Royal Institution in London, in which Michael Faraday did his work. At the outset, however, he placed great emphasis on applied research on stoves and fireplaces. He left his estate to create and endow the American Academy of Arts and Sciences in Massachusetts, which to this day bestows its Rumford medal. If he is not a household name celebrated in the schoolbooks of his native land, that may be because of his checkered political history.

Heat capacity describes the amount of heat (in calories or other units) required to raise the temperature of a unit mass by 1°C. For water just above the freezing point, that is one calorie per gram of water. At its 0°C melting temperature, water absorbs "latent heat" (of 80 calories per gram) without a change in temperature until it melts. It then absorbs the heat necessary to raise its temperature—the sensible heat—until it reaches its 100°C vaporization temperature; thereupon water absorbs latent heat again (540 calories per gram) until it vaporizes.

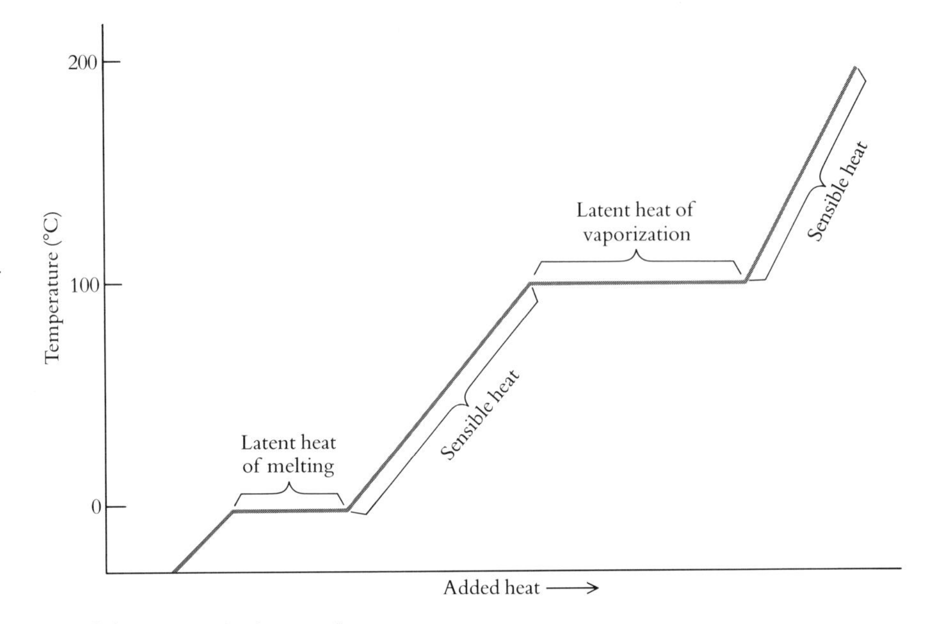

merely a few ounces. The holding capacity of the two containers of water can be said to represent the heat capacity; the depth of the water represents temperature. It is an observed fact that as heat is absorbed, each degree of freedom acquires an equal amount of added energy. This equipartition principle means there is an equal increase in the translational, vibrational, and rotational temperatures in all but the most unusual cases. The temperature then is a measure of the level to which these various modes have together taken up energy.

Another important parameter is the property of heat conductance, or thermal conductivity. This is stated as being the units of heat flowing across a unit cross section per unit time at a unit difference in temperature. As experience invariably shows and the second law of thermodynamics declares, heat always moves from a hot body to the cold. The greater the difference in temperature, the greater the flow. In the candle the flow of heat from higher-temperature molecules to lower-temperature ones through collisions warms the next increment of wax as the wax above melts into the pool.

The heat from the flame that raises the temperature of the wax without melting it is termed sensible heat. It is called "sensible" heat because, in a word, you can feel it. At a critical point the temperature stops rising. The heat capacity at this point increases abruptly, and it takes an extra increment of heat, called the latent heat, to melt the wax. In the familiar case of ice, at 0°C, it takes an input of 80 calories per cubic centimeter of contained water (ice is less dense and so floats in water) to melt it. The latent heat of melting wax is lower.

The heat that flows through a candle flame all originates in the heat of combustion that arises in the reaction zone and from, in lesser quantity, the yellow tongue of the flame as well. Here physics must unite with chemistry, for the heat of combustion is a special case of the heat of chemical reaction.

A molecule is held together by chemical bonds between its atoms. These bonds store potential energy, much as a spring does. A chemical reaction, such as combustion, involves the rearrangement of chemical bonds, the breaking of some, and the formation of others. In the burning of a typical fuel, the bonds between carbon and hydrogen atoms spring apart and new bonds form between these atoms and oxygen atoms. This reaction is exothermic; it releases as heat the excess or difference between the energy stored in the chemical bonds of the hydrocarbon fuel and that stored in the molecular bonds of the end products, carbon dioxide and water. Most chemical reactions are endothermic. They require the input of energy, such as the energy of sunlight (which ultimately synthesizes all fuels), because the bonds in their end-product molecules store more energy than that in the molecules used in making them. To sustain combustion, the innumerable reactions in a candle flame or a fire, which include endothermic as well as exothermic reactions, must in sum be exothermic. Certainly, we find unwanted fires to be exothermic and self-sustaining. We shall see, however, that in the fireplace and in many unwanted fires some materials are consumed by the fire only because fire in surrounding fuels supplies the necessary heat to keep them burning.

The heat of combustion of a fuel is the net energy released as heat by the complete oxidation of a unit mass of the fuel; thus, in the case of the simple hydrocarbon methane:

$$\underset{\text{methane}}{CH_4} + \underset{\text{oxygen}}{2O_2} \rightarrow \underset{\text{carbon dioxide}}{CO_2} + \underset{\text{water}}{2H_2O} + \text{heat}$$

Each gram of methane burned releases 13.2 kilocalories of heat. Candle wax is a mixture of larger molecules such as paraffin, stearic acid, palmitic acid, and other hydrocarbons, each with its own heat of combustion (*see the table on page 88*). While fuels differ considerably in heat of combustion, they all yield about the same amount of heat per unit mass of oxygen consumed in their combustion. Since in many cases burning is limited by the availability of oxygen, this near-uniform behavior facilitates rough calculation of combustion problems.

The combustion reactions in the candle flame do not by any means go in the single step from fuel to end products shown above for the combustion of methane. On the contrary, they proceed through a seeming chaos of intermediate reactions, some of them reversing and repeating and some of them endothermic. These are excited by the presence of a host of intermediate products with fleeting lifetimes, as we shall see when we consider the chemistry of the flame.

Some Heat Capacities

	Specific heat* (cal/g · °C)
Brass	0.07
Asbestos	0.195
Glass	0.161
Paraffin	0.69
Porcelain	0.26
Stearic acid	0.55
Water	1.0
Mercury	0.033
Air	0.24

*Varies with temperature; the values given are for room temperature. Strictly speaking, the specific heat of a substance is the heat capacity relative to that of water at 15°C. The actual value for water is 0.99976 at 15°C.

As is evident, the heat of combustion of the wax in the candle suffices to drive the numerous endothermic processes that keep the candle burning. The heat of combustion supplies enough sensible heat to raise the solid wax in the candle below to the melting point, plus enough latent heat to melt the wax in a pool on top, plus enough sensible heat to bring the liquid wax in the wick to the boiling or vaporization point, plus enough latent heat to vaporize the liquid wax, plus enough sensible heat to raise the temperature of the vapor to the combustion point . . . and so to keep the cycle going.

There is plenty of energy, besides, to emerge as heat and, more productively, as light. That is why the candle is worth burning. The difference between the energy radiated as heat and as light is a matter of wavelength. It was Faraday who first proposed, from an experiment he performed in 1845, that "light itself (including heat and other radiations, if any) is an electromagnetic disturbance in the form of waves." The quotation comes from the historic paper, published 20 years later, in which James Clerk Maxwell constructed Faraday's proposal into a detailed, quantitative electromagnetic theory of light. In answering the "if any" in Maxwell's parenthesis, the electromagnetic spectrum has since been stretched out to ultrashort gamma rays at one end and to the longest radio waves at the other. Heat waves lie in the spectrum just beyond the red long waves of visible light. As the temperature of a solid body increases, the energy it radiates—loses by radiation—goes out in correspondingly shorter wavelengths. The color of its incandescence goes from red toward white. In apparent accordance with this experience, the classical equations of electromagnetics show the intensity of radiation from an ideal "black body" increasing with the frequency or inversely with the wavelength of the radiation. Closer inspection of radiation from solid bodies reveals that the plot of energy against wavelength reaches a maximum intensity, falling off steeply, however, at shorter wavelengths. The required amendment of the classical equations came early in this century with the quantum theory of Max Planck. That theory recognized the particle nature of light along with its wave nature (*see the box on page 22*).

In the candle flame the incandescent particles of soot supply most of the light as they enter into combustion in its yellow tongue. Although most of this radiation occurs in the infrared, or at invisible wavelengths, the small fraction that is emitted in the visible region is sufficient to provide useful illumination. The brightness of the flame is very much a function of the volume of soot particles in it as well as of its temperature. From the soot light is emitted in two kinds of spectra. The incandescent particles, heated to the combustion temperature of carbon, radiate the continuous spectrum of solid bodies. As the carbon atoms go into combustion with oxygen atoms diffusing in from the surrounding air, however, the reactions and the participating atoms and molecules radiate a discontinuous spectrum. The lines and broader bands of heat and light radiation in such spectra bespeak the energies characteristic of the many quan-

Michael Faraday at age 51 in 1842.

Michael Faraday

One of 10 children of a village blacksmith, Michael Faraday was lucky to have a chance to learn the basics of reading, writing, and arithmetic at a day school before starting his working career at age 12. It was a further stroke of luck that he became apprentice to a bookbinder, where his interest was soon drawn to the scientific books passing through the shop. Around 1812, when Faraday was just past 20, a customer gave him tickets for a series of lectures given by the great chemist Humphry Davy at London's Royal Institution. Soon young Faraday was spending his spare moments and pennies on scientific experiments and corresponding with Davy, who had just been knighted for his accomplishments in the study of electricity.

Faraday's laboratory at the Royal Institution in 1870, three years after his death.

Early in 1813, Faraday abandoned his career as bookbinder and, with Davy's help, got a job at the Royal Institution as a laboratory assistant. He soon became Davy's personal assistant, but Davy immediately encouraged him to pursue research projects of his own. For the rest of his career, Faraday was a fully supported resident investigator at the Royal Institution. This was a most fortunate situation, because research in those times was conducted almost solely by men of independent wealth or by those who found wealthy benefactors. The research university and generous government support were far in the future.

Faraday's research extended the discoveries of Davy in many directions. He made his first mark with studies of various inorganic chemical compounds, discovering several new ones, and with his success at liquefying several gases. In the latter research, he became the first to achieve temperatures lower than the zero point of the Fahrenheit temperature scale. Davy felt that Faraday's published reports did not give enough credit to himself for the original work in this research, and bitterness began to grow between the two men. Many later historians have suggested that Davy's resentment grew in large part from his realization that Faraday would surpass the achievements of his benefactor.

Faraday also extended Davy's research on the relation between electric current and chemical reactions. He measured carefully the quantity of electricity involved in various reactions, so that the basic constant used in such calculations today is known as the Faraday constant. He also measured the electric conductivity of various substances.

The work with electrochemistry led Faraday into studies of electricity that we would now classify as research in physics, and it was here that he achieved his most lasting fame. He showed that the movement of a conductor between the poles of a magnet induces a current in the conductor, and that a changing electric current in one wire induces an electric current in an adjacent wire. The powerful concept of the "field" made its first appearance in his efforts to interpret his experimental results.

Faraday established the connection between electromagnetism and light in a highly original experiment he performed in 1845. He showed that a block of glass placed between the poles of an electromagnet becomes capable of rotating the plane of vibrations in a beam of polarized light. This phenomenon, known as the Faraday effect, plays an important role in modern communications devices that use optical fibers.

Faraday's life work gave James Clerk Maxwell his life work. With his superlative mathematical talent, Maxwell constructed the comprehensive statement of the laws of electromagnetism, which are at the heart of modern theory.

Despite his fame as a popular lecturer and research scientist, Faraday tried to keep on working quietly in his laboratory. He refused various honors, including knighthood, and he once flatly refused a request from the government to supervise a project to devise means of using poison gas as a battlefield weapon. His final request, which was honored, was that he be buried under "a gravestone of the most ordinary kind."

tum jumps in energy state that characterize the several chemical species in the gas around the particles. Glowing combustion of solid fuel supplies the enchantment of the fireplace and much of the pyrotechnics one sees in an unwanted fire.

More of the heat and less of the light comes from the primary reaction zone at the bottom and on the surface of the luminous tongue of the flame. The fuel fragments and combustion products each broadcast their identifying wavelengths. The faint blue light radiates principally from two excited molecular fragments, C_2 and CH, produced as short-lived intermediates of intense reactions. The color is faint because only a fraction of the wavelengths are in

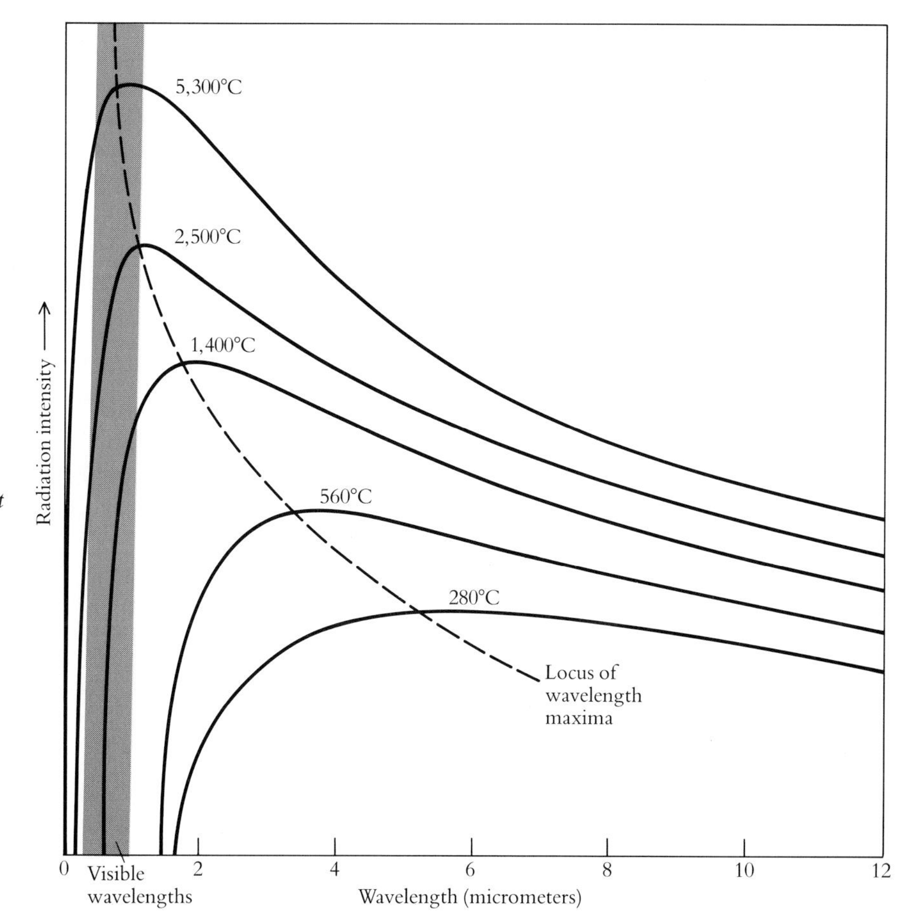

Radiation intensity from a hot body peaks at shorter wavelengths the higher the temperature, as can be seen in the progressive shift in color of an incandescent body from red toward white. Above absolute zero all bodies radiate some energy, all of it at wavelengths longer than the visible, until they approach the temperature of 600°C. This can be seen from the plots of the different distribution of energy by wavelength of an ideal black body at the increasing temperatures shown. The total energy radiated—the area under each of the curves—increases as the fourth power of the temperature.

Black-Body Radiation

A long-standing pursuit of physicists has been an explanation for the behavior of radiation. A particularly difficult case, radiation emitted by hot objects, ultimately produced a revolution in physics.

Isaac Newton thought light rays consisted of particles. This notion, because of Newton's stature, persisted for a century, until Thomas Young showed the wave nature of light. Soon afterward, key experiments in the new sciences of electricity and magnetism by André-Marie Ampère, Michael Faraday, and others led James Clerk Maxwell to formulate the classical theory of electromagnetism, including visible light. Maxwell predicted that oscillating electric charges would emit electromagnetic radiation. Heinrich Hertz, in a series of historic experiments performed from 1885 to 1889, confirmed that prediction. From this work came the whole of electronics technology, including radio and television antennas in which oscillating electrons emit what we call radio waves.

All bodies above absolute zero temperature emit electromagnetic radiation. This arises from the thermal motions of the electrically charged constituent parts of matter (Maxwell's oscillations). The observed dependence of the intensity of this radiation on wavelength is shown in the figure on page 21. The classical theory could not predict the shape of these curves at short wavelengths, stating instead that the intensity increased without limit. Then in 1900 Max Planck made the extraordinary suggestion that the answer lay in considering the radiant energy as occurring in discrete packets, with changes in energy taking place only by gains or losses of complete packets, never by fractions of them. In 1905 Albert Einstein took this idea another step by explaining the energetics of a light beam dislodging an electron from a metal surface: the photoelectric effect. The discrete packets of radiant energy came to be called photons and the new theory, the quantum theory. Despite all this effort to explain radiation, it remains true that electromagnetic radiation has the properties of both particle beams and continuous waves. Conversely, it has been shown that particles such as the electron also behave like waves.

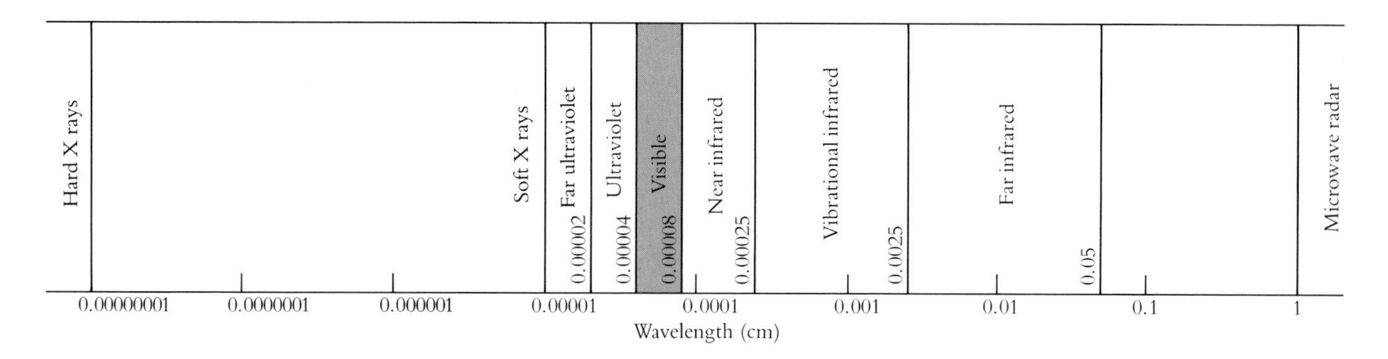

The electromagnetic spectrum is plotted from the centimeter wavelengths of microwave radar down to the 0.00001-centimeter wavelengths of "soft," or relatively long-wavelength, X rays to locate the infrared wavelengths at which molecular vibrations of carbon dioxide molecules radiate energy (see the figure on page 13). That band in the spectrum reaches from 0.0025 centimeter down to the 0.00008-centimeter boundary of the visible spectrum at its red end.

the visible spectrum, and blue because that is how the eye sums up their output. Much other radiation, in the heat or infrared wavelengths outside the visible spectrum, is also emitted from the reaction zone.

The fraction of the total heat of combustion emitted as radiation has recently been measured for large flames of propane. It comes to near 25 percent. The rest goes off in the kinetic energy of the motion of the atoms and molecules within the flame and finally out into the surrounding air, which is slightly warmed.

The radiation is equal in all directions, or isotropic, and so it is possible to calculate roughly the amount of radiation fed back to the candle top. If one makes the convenient assumption that the radiation comes from a single point about two-thirds of the way toward the top of a four-centimeter flame, the solid angle subtended by a two-centimeter candle top is 4 percent. From a conservative heat of combustion of 10 kilocalories per gram of wax, the 4 percent of the 25 percent of the heat available as radiant energy yields 100 calories at the candle top. This is enough to supply the 15 calories per gram required in sensible heat to warm up the wax 30°C to its melting point plus the 40 to 50 calories per gram in latent heat to melt the wax into the pool at the base of the flame. A good deal more heat, 225 calories per gram, is needed to raise the molten wax to its 400°C vaporization temperature and 100 calories per gram more is needed to vaporize it. Since the downward feedback from the flame cannot supply that heat, candle wax cannot sustain its own combustion. At least, not without a boost: a wick. By capillary action the wick draws the molten wax up into lateral exposure to radiation from the blue reaction zone. Here the intense feedback of heat not only vaporizes the wax but pumps in 300 or so additional calories per gram to bring the vapor to its reaction temperature of 1,400°C.

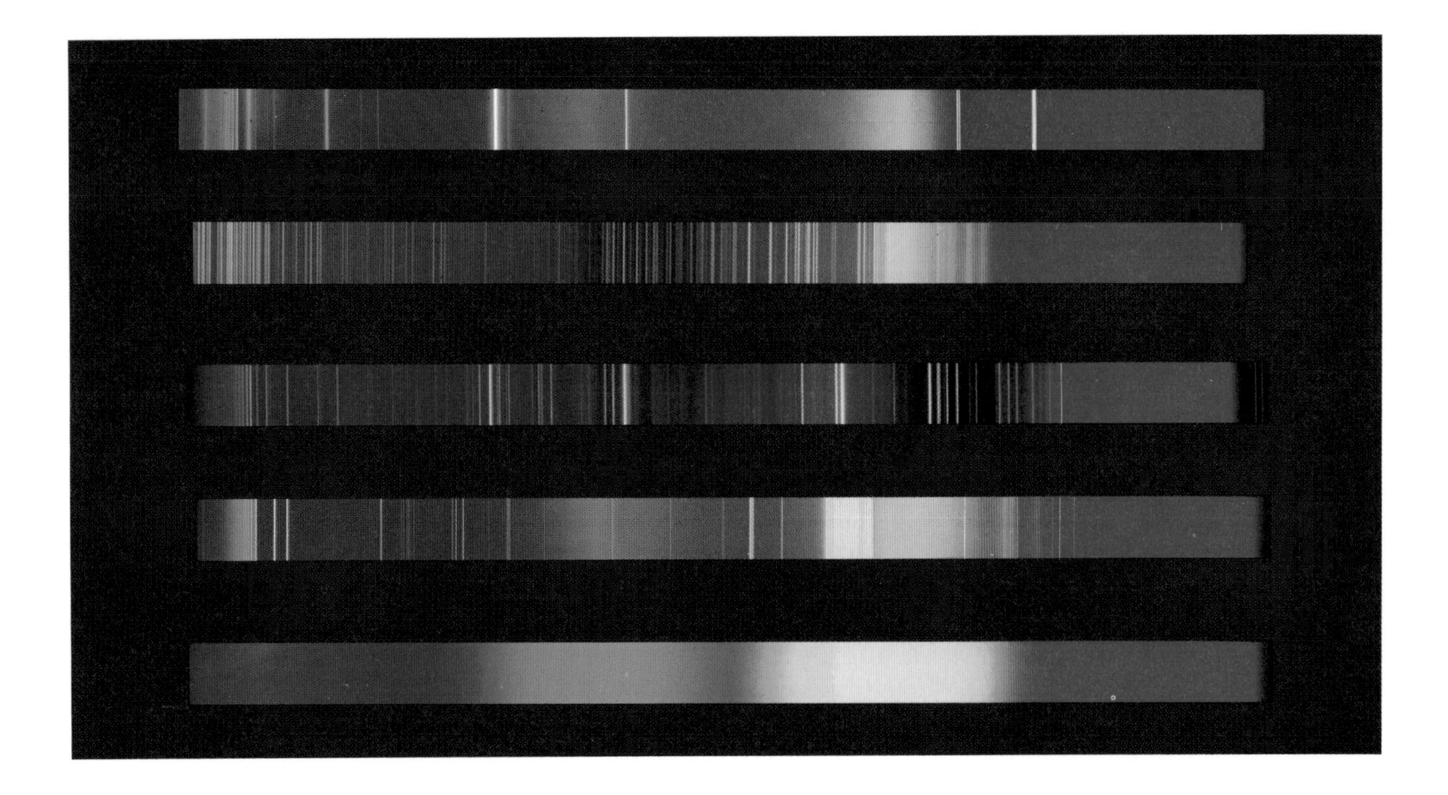

Diffraction, or spreading out, of light by a prism or grating reveals the spectrum of the radiation of which it is composed. Across the spectrum, the energy carried by the photons that registered their presence on the film increases from the red toward the blue. Each line signifies the emission of photons at that wavelength by the quantum jump of electrons from a particular higher to a lower energy state, or orbit. Such line spectra are emitted by individual atoms dissociated from one another in the vapor phase. These spectra, from the top, identify lithium, iron, barium, and calcium. Solid bodies emit continuous spectra like that at the bottom, with the radiation originating not alone from changes in the energy states of individual atoms but also from the associated action of the electronic bonds holding them together in the structure of the solid.

Just as the wick is required to sustain the steady state of a candle, wicks of various design make possible the use of oils as lamp fuels. A variety of ingenious designs for oil lamps have been developed over the centuries, using wicks of various shapes and sizes. Most such lamps provide ways of adjusting the length of the wick to achieve desirable flame characteristics, usually the greatest radiation of light with the least smoke production.

The flame character is affected by the availability of air. Too much wick and too little air result in incomplete combustion and a sooty lamp chimney. Increased air flow may blow the flame off the wick and put it out, or it may increase the mixing of fuel and air, resulting in more intense burning with a hotter flame. Air turbulence may cause flickering and unsteadiness of the flame temperature and luminosity.

The lamp chimney controls the air flow. If you examine a kerosene lamp, you will see that the collar around the wick is fixed to a base with many holes in it so that air can enter from below and be channeled inward and upward to the base of the flame. When the chimney is put in place and convective flow is established by the heat of the flame, this is the only way air can reach the flame. Try the lamp with and without the chimney. With it, you have a bright, steady flame with no sooting; there is enough air for complete combustion. Without it, the flame is less bright and unsteady and will most likely produce visible sooty smoke. The air is not reaching the wick very efficiently. It is almost a secondary effect that the chimney protects the flame from breezes.

Although a small flame cannot be sustained on a small pool of fuel such as kerosene, there is a pool size at which flaming can be sustained without a wick. For the candle, we found that only a small fraction of the radiant energy reaches the candle top; for a large-pool fire, this radiative feedback may approach 50 percent, providing sufficient energy for sustained flaming.

The Chemistry of the Flame

From the physical phenomena we turn now to the chemistry of the candle. The dichotomy of physics and chemistry serves, of course, purely as a convenience, principally for the maintenance of the disciplines and their associated professions. Necessarily, this review of the chemistry of the flame will evoke again the flow of material through the flame propelled by physical forces.

The combustion reaction effects the liberation of the potential energy in the chemical bonds holding together the molecules of the fuel. The candle wax is a fuel. A fuel is any substance that will release energy when it reacts with the oxygen in air, given a start from an external heat source, or igniter. Typical fuels are wood, paper, gasoline, kerosene, animal oils and fats (whale oil, tallow), and other substances derived from plants or animals. These compounds are called organic chemicals because they have their origin in life processes. The organic chemicals all have the element carbon in common; organic chemistry is carbon chemistry.

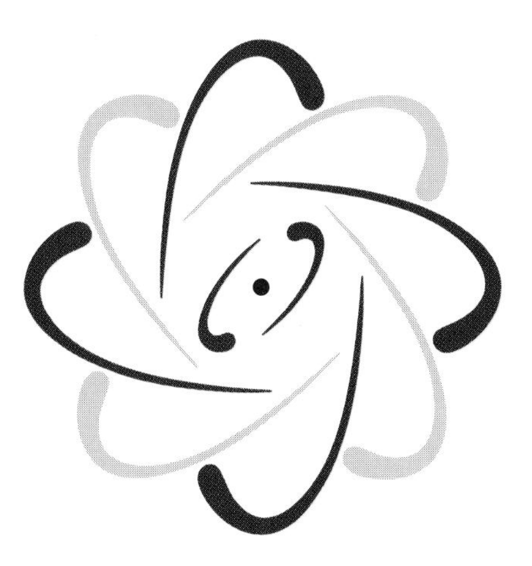

The carbon atom owes its chemical virtuosity to the four electrons that are present (brown) and absent (gray) in its outer orbits. (The nucleus of the atom is the dot in the center of this purely symbolic diagram.) Those outer orbits are "satisfied" by the presence of four additional electrons, matching the structure of the atom of the noble (chemically inert) gas neon. In its four half-empty outer orbits carbon can capture the electrons of two to four other (and different) atoms in triple, double, or single bonds, forming stable compounds.

Carbon is a versatile element. Like other atoms, the carbon atom has a nucleus and a group of electrons orbiting about the nucleus. The nucleus of any atom makes up most of its mass but occupies a very small space. The electrons swarm about the nucleus at a distance, and the space occupied by these circling electrons defines the space occupied by the atom. A typical atom has a radius of about 10^{-10} meter (0.1 nanometer, one angstrom, 4×10^{-9} inches, or four-billionths of an inch). A rough estimate shows that magnification of the nucleus to the size of a grapefruit would place the innermost pair of electrons orbiting at a distance of a mile.

Carbon chemistry can be said to be governed by the presence or the absence of four electrons in the four outermost orbits. In these orbits there are four electrons, but there is room for four more. The carbon atom thus readily participates in chemical reactions in which other atoms provide the four additional electrons to fill those outer electron orbits. Because each of the four electrons may come from a different element, the possible combinations of elements with carbon are in effect endless.

Principal candle fuels are the paraffins or waxes, in which carbon atoms linked to one another by single bonds form their two other possible bonds with hydrogen:

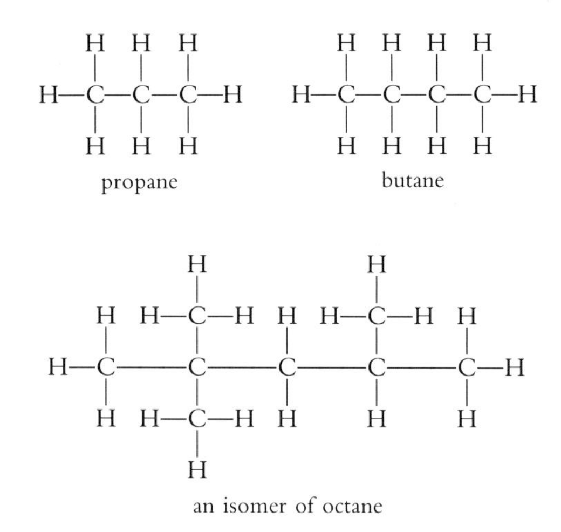

propane butane

an isomer of octane

The smaller paraffin molecules shown here form gases or liquids at room temperature; larger ones form waxy solids. The paraffin molecules in candle wax contain about 18 to 20 carbon atoms.

Oxygen appears in many common fuels, particularly as components of groups:

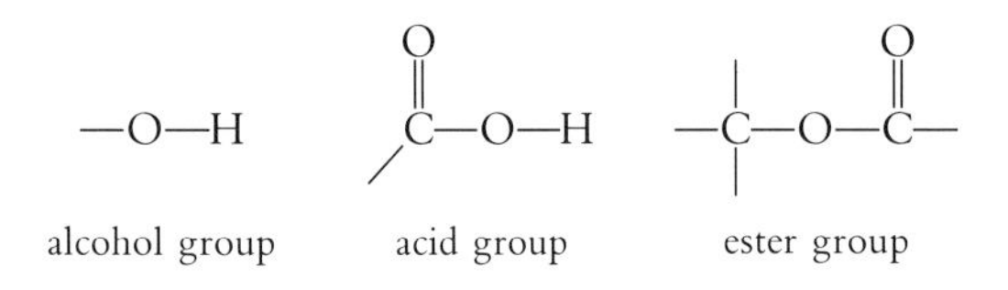

alcohol group acid group ester group

Most candle waxes contain fatty acids, such as:

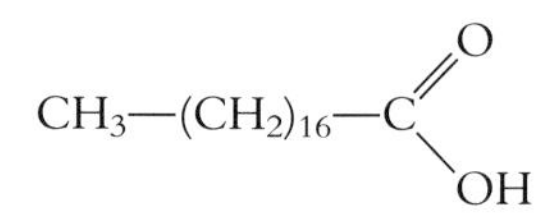

Wood supplies cellulose as a principal fuel molecule, shown here in a schematic diagram that delineates the backbone structure:

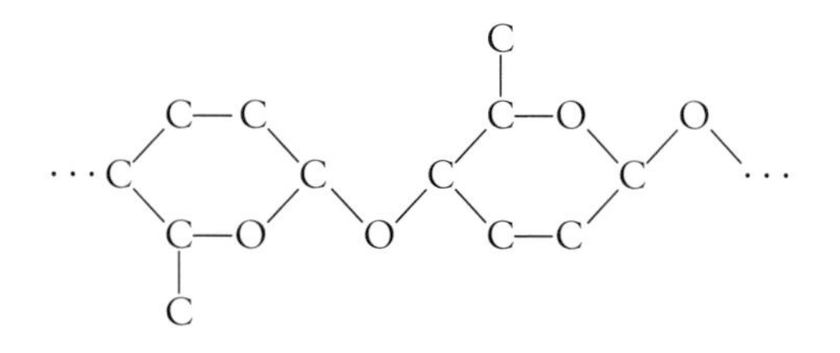

The candle-wax paraffins and fatty acids melt at about 45 to 62°C. Not much heat is required: only 40 to 50 calories for a gram, compared with the 80 calories needed to melt a gram of ice. As the liquid wax ascends the wick and absorbs the heat cascading in from the reaction zone close by, its temperature rises. Even before it takes in enough latent heat to vaporize, some shifting back and forth of chemical bonds begins and the structure of the molecules changes. A single molecule may undergo isomerization, or a change in its structure without a change in its composition, thus:

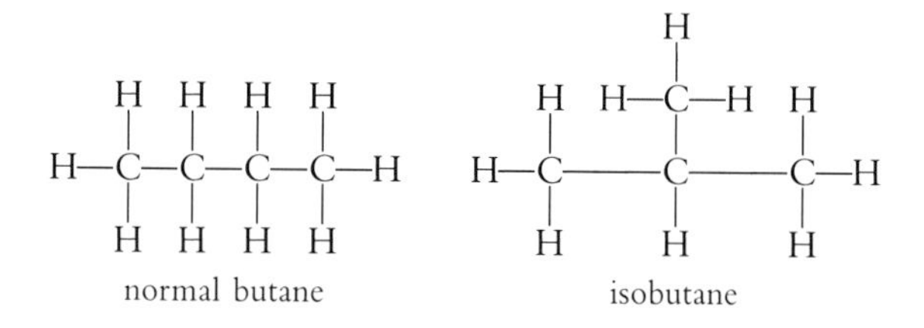

normal butane isobutane

Exchange of parts between molecules may also begin to occur in liquid wax:

$$\mathrm{R{-}X} + \mathrm{R'{-}Y} \rightleftharpoons \mathrm{R{-}Y} + \mathrm{R'{-}X}$$

Recent studies have demonstrated that such exchanges and rearrangements occur in liquids at fairly low temperatures, as low as 120°C. For candle wax these reactions proceed at higher temperatures, around 300 to 400°C. We can imagine the candle wick as a very busy place, with rapid molecular motions and many collisions that involve occasional trading of parts and rearrangements of structure among the molecules of the liquid wax.

At some point as the wax climbs upward in the candle wick, the temperature reaches the vaporization temperature of the wax. This is analogous to the vaporization point of water (100°C, or 212°F), but the temperature is much higher, about 350 to 400°C. The wax molecules and fragments of molecules then separate from the liquid surface and commence moving about as a gas. Immediately around the wick, where the flame is almost transparent, the molecules and fragments gush outward. Combustion chemists refer to this region as beyond the fuel-rich limit—that is, there is not enough oxygen present to sustain flaming combustion.

The consumption of oxygen in the combustion zone creates a strong chemical gradient, causing more oxygen to diffuse from the air into the combustion zone. It does so against the outward pressure of thermal diffusion, a demonstration of the force of chemical diffusion. The chemical diffusion effect gives the candle flame its technical name: it is a diffusion flame. In contrast, in the Bunsen burner and the kitchen gas range air, and so oxygen, are entrained in the stream of fuel before ignition in the flame; hence they are called premixed flames.

The chemical reaction of combustion can occur when fuel and oxygen molecules mix only if the temperature is high enough. For wax, as for other materials, there is a critical temperature at which it sustains combustion with no additional external energy input, and there is another, lower critical temperature at which it sustains combustion provided an outside initiator is present. The initiator in the candle, always present, is the steady-state flame itself. It is fed by molecules in the vapor generated in the immediate previous instant of its life, starting from the moment it was ignited by a match.

At the critical temperature, the combustion reaction proceeds thus:

$$\underset{\text{wax fragment}}{-(CH_2)-} + \underset{\text{oxygen}}{\tfrac{3}{2}O_2} \rightarrow \underset{\text{carbon dioxide}}{CO_2} + \underset{\text{water vapor}}{H_2O} + \text{heat}$$

This representation tells nothing about the details of the chemical reaction—about what collisions occur, what intermediate species are produced, and so on. It does not explain how it was that when we placed a cold spoon in the top of the flame just inside the yellow region, the spoon was quickly covered with soot. Evidently soot particles abound. Thus we find, as Faraday did, that the combustion reactions are complex—that combustion proceeds from a fuel-rich region where carbon is produced and then to fuller burning that produces the end products.

What apparently occurs in the candle flame is first a thermal "cracking" of the wax molecules, breaking them down into fragments. Predominant among these fragments are free radicals, ordinary molecules or molecular groups made highly reactive by the loss of an electron in an outer orbit, leav-

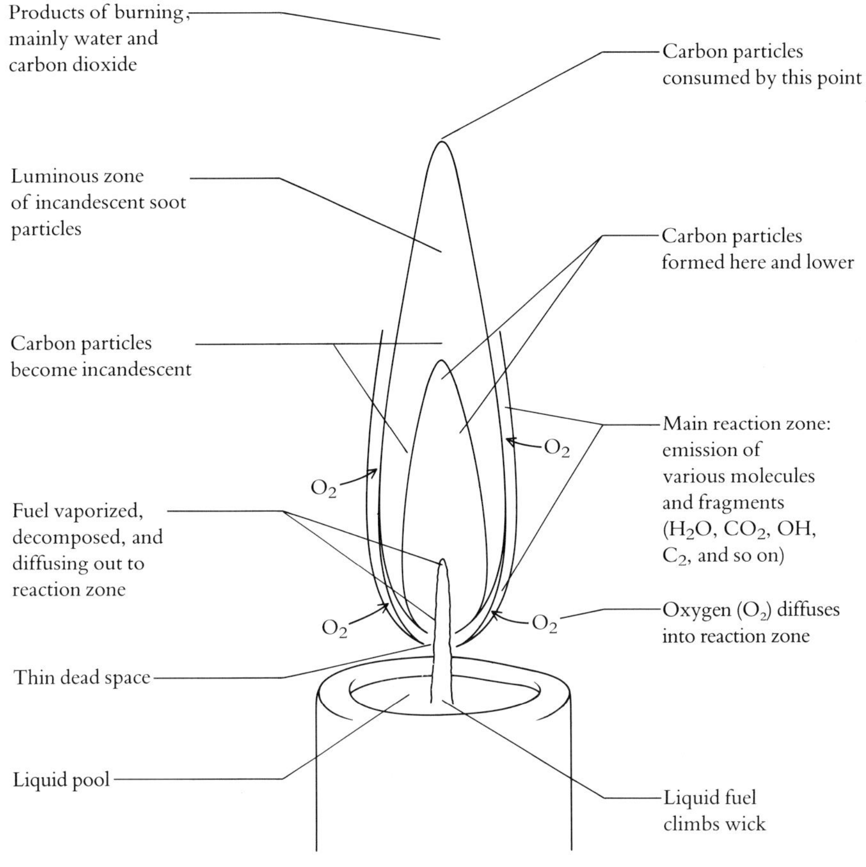

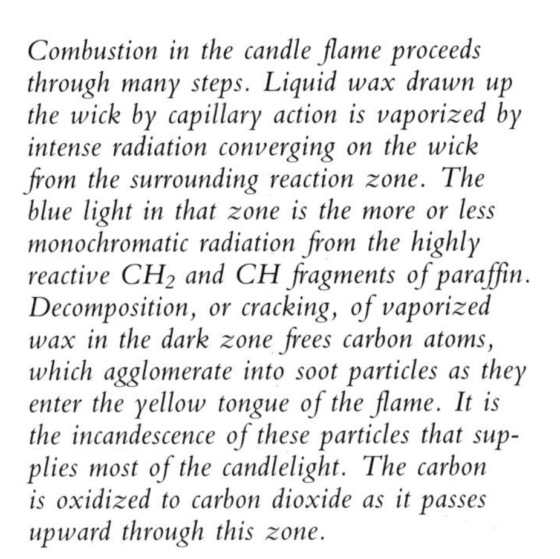

Combustion in the candle flame proceeds through many steps. Liquid wax drawn up the wick by capillary action is vaporized by intense radiation converging on the wick from the surrounding reaction zone. The blue light in that zone is the more or less monochromatic radiation from the highly reactive CH_2 and CH fragments of paraffin. Decomposition, or cracking, of vaporized wax in the dark zone frees carbon atoms, which agglomerate into soot particles as they enter the yellow tongue of the flame. It is the incandescence of these particles that supplies most of the candlelight. The carbon is oxidized to carbon dioxide as it passes upward through this zone.

ing a single, unpaired electron behind. They will react with one another in a terminal reaction or with other chemical species in chain reactions. Oxygen and water molecules, transformed into this free-radical state, supply important "hot spots" that drive the flame reaction. When oxygen is in short supply, as in the dark portion of the flame, the chemistry is dominated by excited fragments of the wax molecules. In this region, atoms are hot and labile, and reactions occur without oxidation, freeing up small carbon-rich molecules, which aggregate into soot particles:

$$\underbrace{-\mathrm{CH_2}\cdot + -\mathrm{CH_2}\cdot}_{\text{two radicals}} \rightarrow \underbrace{\frac{1}{x}(\mathrm{C{-}C})_x}_{\text{soot}} + 2\mathrm{H}\cdot + \mathrm{H_2} + \cdots$$

As soot moves upward in the flame, its incandescence supplies the luminous yellow light. Glowing combustion occurs on the surface of each soot particle:

$$C(solid) + O_2(gas) \rightarrow CO_2(gas) + \text{heat}$$

This oxidation reaction incorporates carbon atoms from the solid carbon particles into molecules of carbon dioxide gas.

Recent studies show that the soot particles become larger as they move up in the flame. Tiny, spherical particles tend to aggregate into chains or clusters. The soot is not pure carbon; hydrogen atoms remain bonded to some of the carbon atoms. The carbon atoms are linked in a planar (two-dimensional) structure similar to that of graphite:

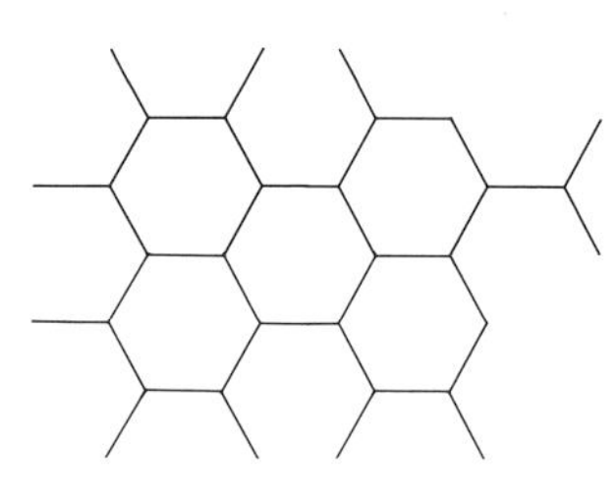

Each straight-line segment represents a bond between two carbon atoms.

Within the upper part of the flame, fragments of fuel molecules continue to be broken down by reactions with free radicals. A free radical can react with a large molecule to "chip away" a part of it:

$$O_2 + H\cdot \rightarrow OH\cdot + O\cdot$$

$$OH\cdot + RCH_2CH_2CH_3 \rightarrow RCH_2CH_2CH_2\cdot + H_2O$$

$$R_nCH_2CH_2CH_2\cdot \rightarrow R_{n-1}CH{=}CH_2 + CH_3\cdot$$

(Here n stands for any number.) Through a complicated family of reactions, the large carbon fragments are eventually oxidized to carbon dioxide, releasing more free radicals to carry on the chain reaction. Among the intermediate products of the reaction series are the methyl radical, $CH_3\cdot$, and the gases ethylene ($CH_2{=}CH_2$) and carbon monoxide (CO). The final steps might look like this:

$$CH_2{=}CH_2 + O_2 \rightarrow 2CH_2O$$

$$CH_2O + OH\cdot \rightarrow CO + H_2O + H\cdot$$

$$CO + OH\cdot \rightarrow CO_2 + H\cdot$$

The presence of all of these species in the candle flame has been confirmed by spectroscopic studies. Thus the conversion of fuel into CO_2 and H_2O is a

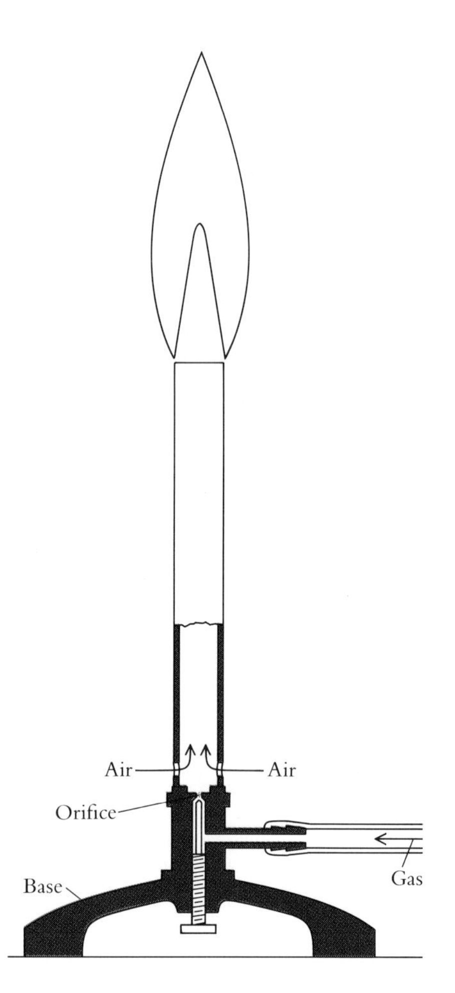

The Bunsen burner has a premixed flame. In contrast to the candle, in which the wax fuel and oxygen in the air find their way into contact with each other by diffusion into the flame reaction zone, the fuel and air in a Bunsen burner are mixed before they leave the burner tube. This stratagem permits admission of oxygen into the flame in precisely the amount that secures complete combustion and so the hottest flame from that fuel.

complicated process that takes place by at least two major paths: by the free-radical reactions in the hot gases and by glowing combustion at the surface of soot particles.

The chemistry of flames is an active field of research, employing instrumentation that "stops" action in the nanosecond (10^{-9} second) range in order to identify and even trap intermediates with fleeting half-lives. Such studies lend precision to the control of reactions in the chemical-process industries and bring new reactions into chemical technology.

A familiar product of flame chemistry is carbon black. To manufacture carbon black, fuel oil is burned with a reduced air supply; a condensing surface traps the soot particles (the cold spoon in the candle flame!). The end products, lampblack, channel black, and furnace black, have many commercial uses: as fillers in the rubber in tires, giving tires their color; as the base for the electrolyte paste in dry cells; as a coloring agent in inks. Control of the glowing-combustion process is important for the manufacture of blacks, as well as for the limitation of the soot produced in flames used for heating and illumination.

Understanding of the free-radical reactions has led to the development of flame poisons, or inhibitors. Such materials, when vaporized, produce radicals that attract the $H\cdot$ and $OH\cdot$ radicals more strongly than do the radicals produced by fuel breakdown. Because the free radicals are trapped by the inhibitor, the combustion reactions slow down for lack of the needed free radicals. This in turn reduces the feedback of radiative energy, causing the flames either to go out or to burn very weakly and slowly. Halogen radicals in the vapor are very effective at trapping $H\cdot$ radicals. Additives containing halogens such as chlorine and bromine are members of a class of chemicals called fire or flame retardants.

Flows of Heat and Mass

Consider the course of a single molecule of wax on its way to combustion in the candle. Initially, at room temperature, the molecule is constrained within the solid phase. As the candle burns downward and the flame approaches our molecule, it receives heat by conduction from above. With the increase in temperature, our molecule vibrates more and more vigorously about its position in the structure of the solid. At last it breaks loose and begins moving about among neighboring molecules, forming weak and temporary relations with them but having no fixed position in a long-range structure. Our molecule is now a part of the pool of liquid wax at the top of the candle.

As it receives more energy, mostly by radiation from the flame above, it moves more and more rapidly. Eventually it is drawn into the wick and up-

Chemical reactions in a flame are studied "noninvasively" with a laser beam that does not disturb the flow pattern of the flame gases (as Faraday's silver spoon in the candle flame did). In this premixed flame of methane gas the laser beam excites a green fluorescence (not visible in the photograph) in large, ring-shaped, or "aromatic," molecules. The subject of the study, at the National Bureau of Standards, is the formation of soot in a flame.

ward by capillary attraction in the interstices of the wick fibers. As it moves upward in the wick, its motions become ever more rapid until it breaks loose from all constraints and flies into space. It is now in the vapor phase.

Absorbing radiant energy and colliding with energetic neighbors, our molecule acquires still greater velocity and greater internal energy (rotation and vibration). Overall, its random motions tend to carry it outward, away from the wick, until it enters the reaction zone, where abundant free radicals have been produced by interactions of incoming oxygen and outgoing fuel molecules. Through collisions with free radicals, our molecule is gradually broken apart, releasing energy. This combustion process occurs very rapidly and over a very short distance. The chemical-bond energy released is in part transformed into the diffusion and convection motion of the product molecules, in part conducted back toward the wick by rapid collisions with molecules behind, and in part radiated in all directions.

In the oxygen-poor lower region of the flame, however, some of the carbon in our molecule escapes combustion. Aggregating in incandescent soot particles, the carbon ascends into the upper tongue of the flame, to which it lends its luminance. The carbon encounters oxygen diffusing inward and undergoes glowing combustion. The components of our original molecule thus eventually leave the flame as carbon dioxide gas and water vapor, streaming into and warming the air above, having released much energy in the burning process.

Carbon Chemistry

The carbon atom owes its chemical versatility to the electrons in its four outermost orbits. There are four electrons in these orbits, but there is room for four more (each orbit will hold two electrons). The carbon atom readily participates in chemical reactions that provide the four additional electrons to its outer electron orbits. The combinations carbon can make with other elements are almost endless—with the 92 naturally occurring elements, nearly 72,000,000—because each of the four electrons may come from a different element. Thus carbon can combine to form:

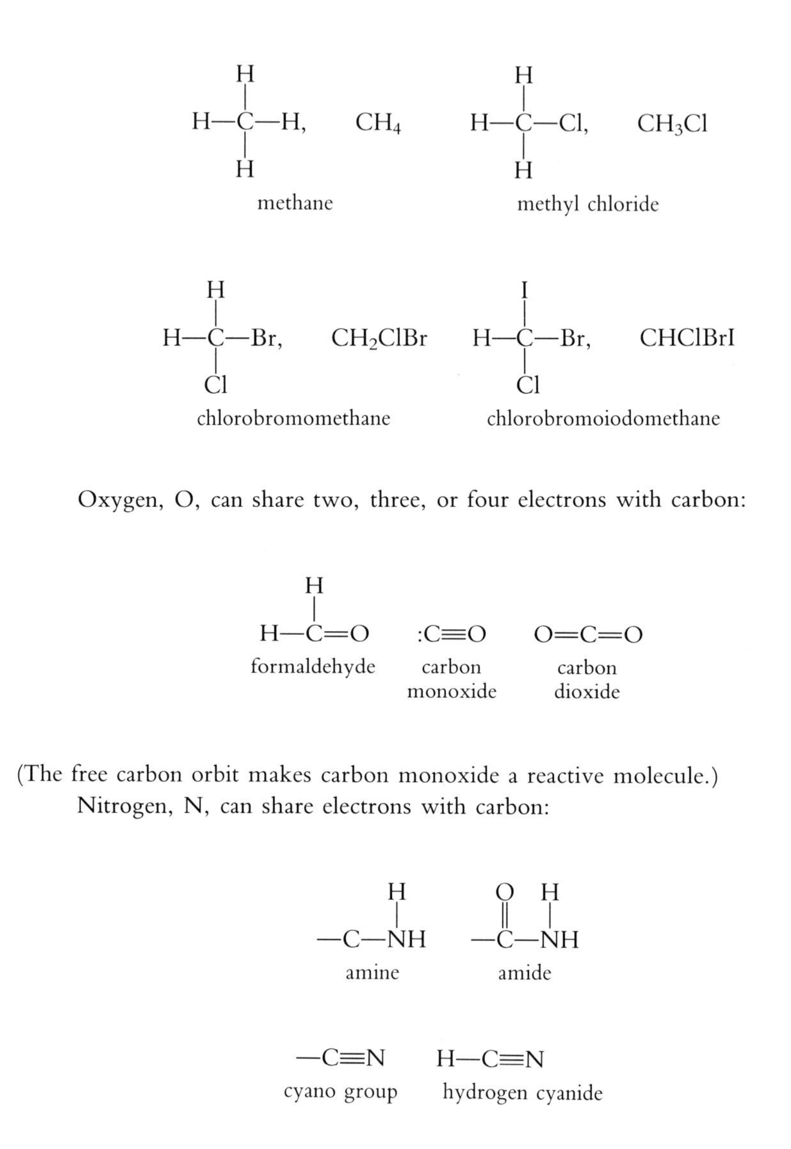

Oxygen, O, can share two, three, or four electrons with carbon:

(The free carbon orbit makes carbon monoxide a reactive molecule.)

Nitrogen, N, can share electrons with carbon:

Carbon may also acquire electrons from other carbon atoms by sharing:

—C— + —C— → —C—C—

carbon-to-carbon bond

Sometimes this is double sharing or even triple:

$$-\mathrm{C}- \; + \; -\mathrm{C}- \; \rightarrow \; -\mathrm{C}{=}\mathrm{C}-$$

ethylene group

$$-\mathrm{C}- \; + \; -\mathrm{C}- \; \rightarrow \; -\mathrm{C}{\equiv}\mathrm{C}-$$

acetylene group

The linkages can go on at length in one, two, or even three dimensions:

$$-\mathrm{C}-\mathrm{C}-\mathrm{C}-\mathrm{C} \cdots \mathrm{C}-\mathrm{C}-\mathrm{C}-$$

linear chain

Carbon may form two-dimensional structures:

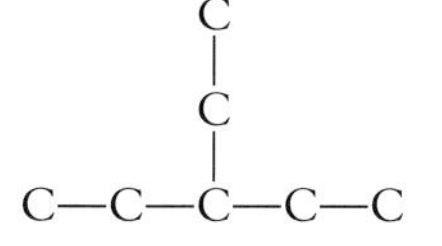

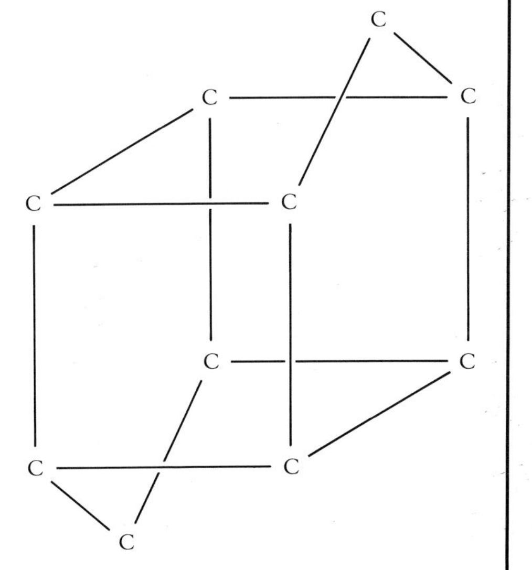

and three-dimensional structures (left).

Here are three fuels of the simplest kind, all-carbon chains (or polymers) saturated with hydrogen:

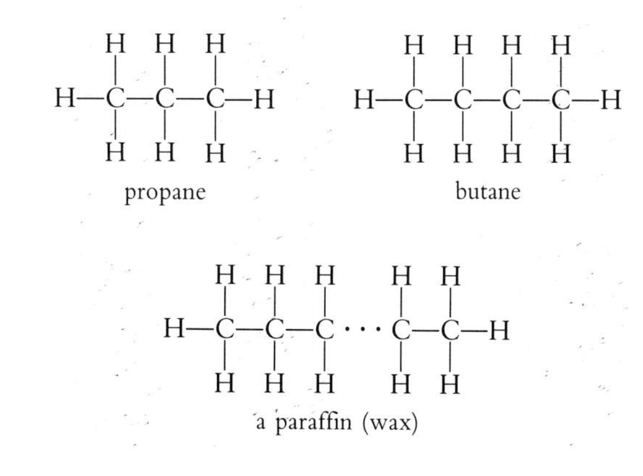

propane butane

a paraffin (wax)

The carbon in these chains can exchange some H for O, two for one.

Radiation conveys the aesthetic pleasure and all of the heat a fireplace gives to the room.

3 *Fire on the Hearth*

To see what it could tell us about the physics and chemistry of fire, we have considered the flame of the candle burning in a steady state on a single fuel. We consider now the unsteady state of fire in a fireplace. Many flickering flames burn here on more than one piece of fuel, often of more than one kind. The difficulty some people experience in trying to get a fire started in a fireplace may make us wonder how it is that unwanted fires seem to start so easily. The fireplace fire gives us nonetheless a model of the scenario and the larger-scale dynamics of an unwanted fire.

Several new factors, not encountered in the burning of a candle, now come forward. First, it is plain that fireplace fuel does not melt. The properties that govern the raising of the fuel temperature to the combustion point come next. The third factor is the effect that fire on one piece of fuel has on a piece adjacent. The fourth is the effect of the noncombustible immediate surroundings of the fire, the fireplace and its design. Finally there is the effect of air movement into and through the fire.

The match flame that lights a candle cannot start a fire on a log. The solid wood does not melt, vaporize, and burn. Even if a flame starts up on the surface, the wood does not sustain it. A green, wet log presents the extreme case. Exposed to prolonged heating on one surface, it will absorb the heat without much rise in temperature. That is because of the high heat capacity of the water in the wood; it takes a whole calorie to raise one gram one degree. Thanks to the high heat conductivity of water, on the other hand, the wet log heats through rather than heating up at the surface. Finally, with its high density it responds slowly overall to a given flux of energy. A block of low-density foamed insulation—made of wood cellulose or plastic—behaves quite differently. Exposed to the same energy flux, it heats up rapidly at the surface owing to its comparatively low heat capacity. Its low heat conductivity and low density keep the incident energy piling up at the surface while the interior remains cool. Even with its surface aflame, the block of insulation will remain cooler inside than the wood.

Plainly, the fireplace wants a supply of dry logs. To overcome the effect of thermal conductivity, paper and kindling offer an increase in the surface-to-

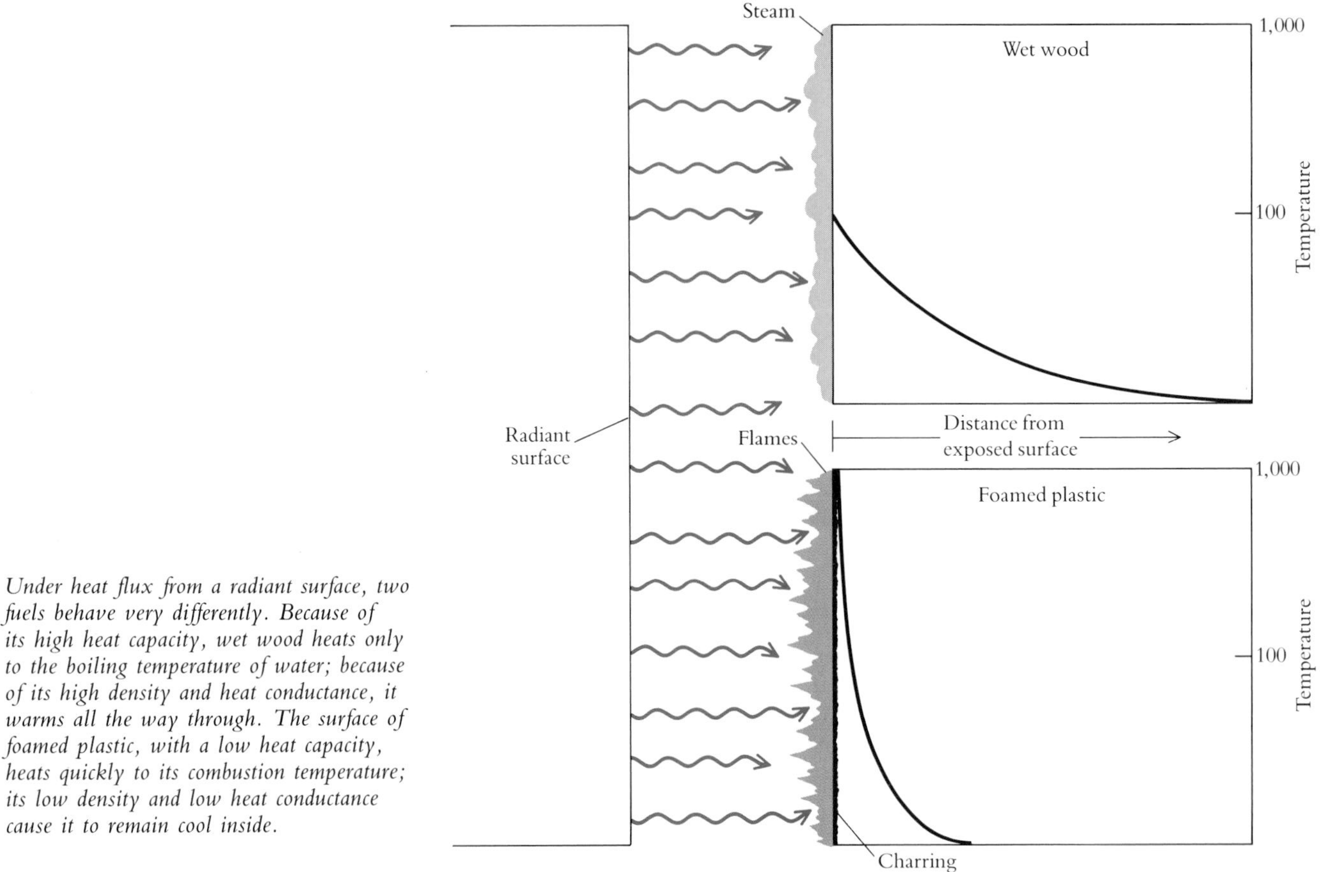

Under heat flux from a radiant surface, two fuels behave very differently. Because of its high heat capacity, wet wood heats only to the boiling temperature of water; because of its high density and heat conductance, it warms all the way through. The surface of foamed plastic, with a low heat capacity, heats quickly to its combustion temperature; its low density and low heat conductance cause it to remain cool inside.

volume ratio. That is the ratio between the surface to which the match is applied to the volume into which the heat is to be conducted. Dry paper, being in essence all surface with low heat capacity, ignites readily and sustains combustion. It sustains combustion long enough, anyway, to ignite the kindling, with its still relatively high surface-to-volume ratio. The kindling burns long enough to get the logs burning, providing the fire has been properly laid.

Wood does not melt when heated; rather, it blackens, chars, and more or less retains its shape until it is burned through. What happens in the charring is that the molecules of cellulose and lignin in the wood undergo thermal decomposition and become fuel gas without passing through an observable liquid phase. This process has the formal name of pyrolysis: decomposition by fire. Softwoods have a lower density than hardwoods and, being made essentially of the same cellulose and lignin, less total heat content, meaning substance

Heat Capacity and Thermal Conductivity

Substance	Heat capacity (cal/g · °C)	Thermal conductivity (cal/h · cm · °C)*
Water (at room temperature)	1.0	5.2
Dry oak wood	0.5	1.5
Water vapor (room temperature)	0.5	0.2
Air (room temperature)	0.24	0.2

*For convenience units are in hours instead of the usual seconds.

available for transformation by pyrolysis and combustion into heat. In consequence, softwoods burn more easily, but do not last as long.

Heating increases the volume of the gas from pyrolysis inside the log, producing cracks in the wood through which the gas escapes to fuel the flames. The combustion of this gas heats the carbon in the char to its combustion point. From the softwoods, especially, vapor from volatile gums sustains bright flames at fissures here and there in the char; enough gas may be contained behind one or another of these fissures to fuel a hissing flame for several seconds or more. On occasion no cracks or fissures have formed to permit escape of the gas; the surface of the log then pops, and the "sparking" fuel tosses a coal or two out on the floor and perhaps on a rug beyond the hearth.

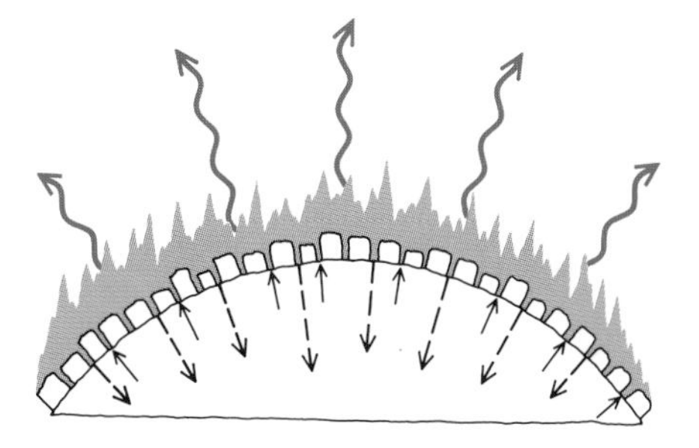

Heat at the surface of a burning log moves inward and outward. Heat moves inward by conduction (dashed lines), decomposing the lignin and cellulose of the wood into gaseous fuel that leaks out (short solid arrows) to sustain the flame. Heat moves outward by conduction, diffusion, and convection in the flames as well as by radiation (wavy lines) from the flames and from the hot surface.

Flames in the fireplace are brightened by the incandescence and combustion of particles of soot, as in the luminous candle flame. Otherwise providing more heat than light, the gaseous combustion reactions in these flames also sustain glowing combustion of the carbon from the solid state at the incandescent surface of the ruddy-glowing char. Spectroscopic analysis would show, in parallel with the candle, line and band spectra from the combustion reactions and a bright continuous spectrum from the incandescent char and particles of soot in the flames.

A log burning vigorously in a fire will not usually continue to burn if removed from it. That is because the layers below the surface of the log will not remain hot enough to produce the gaseous fuel needed to sustain the flame. The heat generated by the fire at the surface is lost by radiation to the sur-

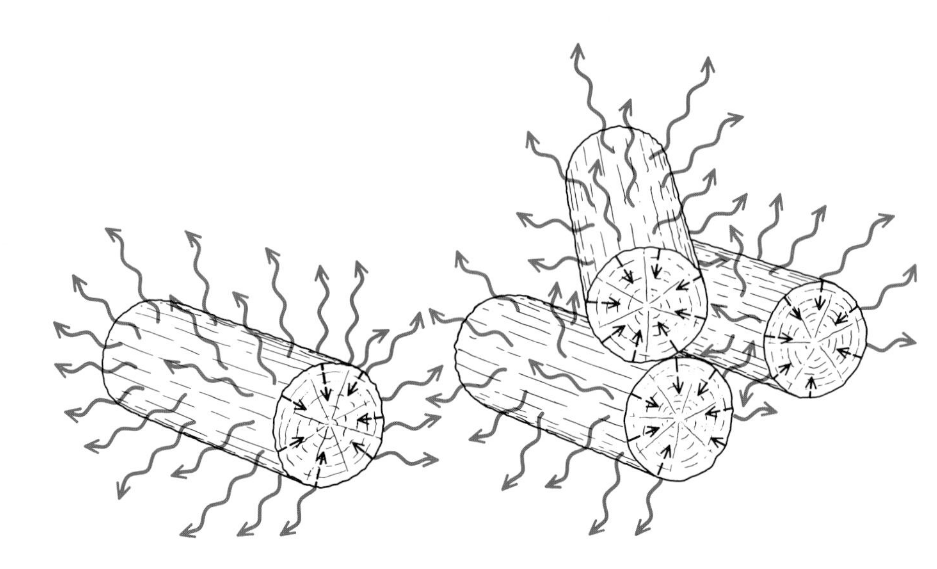

Due to the loss of heat outward by radiation, the surface of a single log soon cools below its ignition temperature and the flames go out. Two or three burning logs abutted to one another sustain their combustion by capturing each other's radiant energy.

roundings and by conduction to the interior. Unless the fire is sustained on the surface, heat conduction ceases to carry enough heat to the interior to bring fuel gas to the surface. Since conduction of heat into the deep layers proceeds as well in the log on the fire and in the isolated log, it must be the loss of heat by radiation from the isolated log that makes the difference. In the log on the fire that heat is not lost. It heats the log facing it, sustaining the fire in that log; that fire generates radiation in return to keep the first one burning. The same effect can be achieved, with a good deal more trouble but to prove the point, by furnishing an isolated log with a curved, well-polished reflector to redirect its radiant energy back to its surface. In the fireplace, the desired result is best secured by adding a third log to reinforce the mutual feedback of radiant energy and thus sustain the burning.

The mass and heat flow in the fireplace is simple. The wood is converted into gases, which go up the chimney, and ashes, which drop to the hearth. The chemical changes, overall, are the oxidation of the wood ultimately to carbon dioxide and water. The heat flow is more complex. The convective flow is up the chimney. Unless the fireplace is smoking, there is no convective heat flow back to the room. Conduction of heat is important to heating up the logs. The radiative flow of heat sustains by mutual feedback the burning of the logs and delivers most of the heat to the room. Radiation to the room comes directly from the flames and glowing char on the logs, from the hot ashes, and from the walls and back of the fireplace.

The immediate surroundings of a fire are important in that, properly designed, they can substantially enhance burning. Just as a polished reflector can direct lost energy back to the fuel, so can the interior surfaces of the fireplace or stove. These walls are not good reflectors; often they are blackened by soot. As they are heated by the fire, however, they reradiate energy and some of this is returned to the fuel.

The ashes beneath the fire become very hot and reradiate intensely. Much of this energy is returned to the fuel. A thick bed of ashes has an insulating effect on the floor of the hearth and also brings the radiating surface of the ashes closer to the burning logs. It is best to always leave enough ashes in the fireplace to take advantage of this radiative feedback. There should be only enough room between the ashes and fuel to provide a comfortable channel for air flow, perhaps half an inch or so. A country hearth will usually be found to be liberally supplied with ashes, not for lack of tidiness on the part of the housekeeper but for providing a warmer, longer-lasting fire.

There has been some quantitative study of the effect of boundaries on the rate of burning. As compared with a sample of fuel burning in the open, researchers have seen increases in fire intensity, as measured by burning rate, of as much as a factor of 2 on placing the fire in an enclosure. The effect is largely one of enhancement of the temperature at the fire by reradiation from the surroundings.

Combustion requires air. Fuel surfaces must be accessible to air or the fire will go out. There is, therefore, an optimum spacing of the logs on the grate that provides plenty of air while at the same time keeping the logs close enough to the ashes and to one another to reinforce burning by radiation. The closer the fireplace boundaries, the larger the role they will play and the less critical will be the log stacking. In a large fireplace with a small fire the boundary effect is small and the stacking of the fuel is more important. In a closed stove, with maximum loading, the stacking is less critical because wall effects are dominant.

As air passes through the fire some of the oxygen is consumed and combustion products such as carbon dioxide and water vapor are produced. The gas stream is heated, expands, and because it is now lighter than the ambient air, it rises. This buoyant effect produces a plume of hot air rising above the fire. As it rises it draws in behind it cool ambient air, which provides a fresh supply of oxygen to the fire. This flow of air is critical for success; otherwise, the fire would quickly be starved of oxygen. The chimney channels the hot gases out of the building. An open hole above, as in the tepee or igloo, would serve the same purpose. The chimney enhances the draft on cold days because of the temperature difference between indoors and outdoors, enough to keep the draft flowing up the chimney with no fire on the hearth. A damper is provided to choke this off. A very serious problem with the fireplace is that the convective buoyancy moves more air than is needed for combustion, and large volumes of warm room air go up the chimney. Much of the design of stoves and of fireplace appliances has been aimed at controlling and limiting this convection.

A kitchen fireplace did the cooking as well as the heating. Pots were hung from a crane attached to the wall and a spit could be mounted in front. The oven at right, with a flue connecting to the chimney, was heated by its own fire; with that fire swept out, the food was roasted or baked by reradiation of heat from the interior surfaces. Such ovens were replaced by iron stove-ovens.

When the first settlers came to America they built fireplaces big enough to walk into, with wide, straight chimneys generally fitted with transverse rods for hanging things that were to be dried or smoked. The fire tender could walk in under the lintel across the opening. The chimney usually kept down the level of smoke inside the room, though there were occasional fluctuations in outside air pressure that caused some smoking. With a good draft, on the other hand, warm air from the room went up the chimney along with the smoke. Additional heat was lost from conduction through the back of the fireplace when that was in the outside wall of the building, as it often was.

The downdraft from outside air pressure fluctuations was subsequently tempered by the introduction of the smoke shelf. A smoke shelf is obtained by offsetting the path of the chimney flue from the fireplace throat. When a puff of air comes down the chimney, it strikes the shelf, is turned by it, and rejoins the upward flow of hot air from the fire. The disturbing effect of air pressure fluctuations on the air flow in the fireplace proper is reduced, and the escape of smoke back into the room is lessened if not eliminated. The damper, added to the smoke shelf, provided a means of extending the shelf and a way of closing off the chimney altogether.

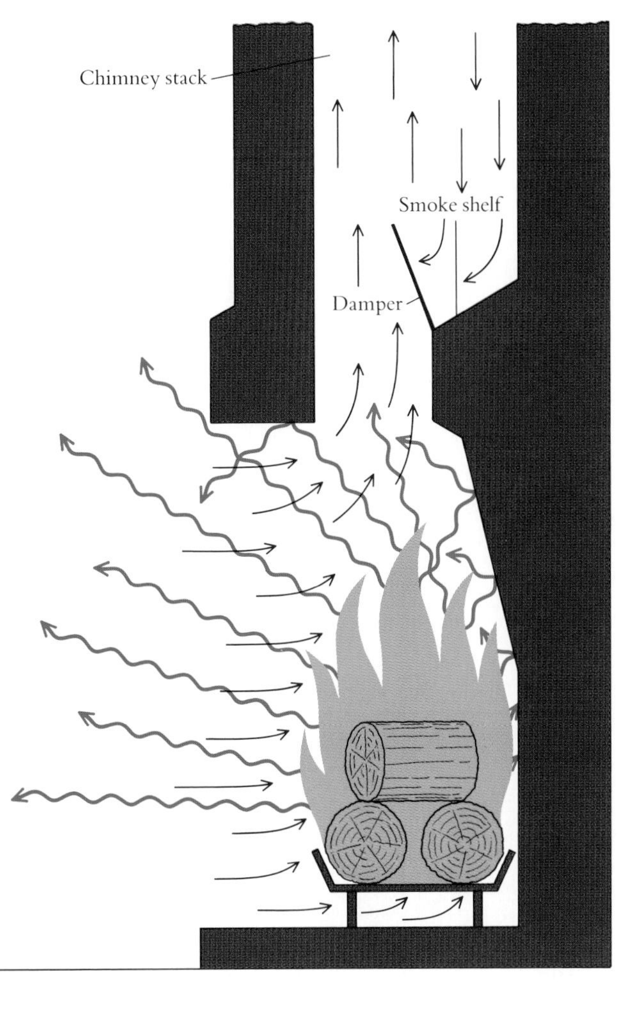

A fireplace is an inefficient generator of heat from fuel. From the approximately 25 percent of the total heat output that goes out in radiation, the room receives only the portion that is radiated into the quadrant in front of the fireplace. Most of the 75 percent balance of the heat of the fire in the combustion gases goes by convection up the chimney. The flow of those gases up the chimney draws the warm air from the room up the chimney with it.

To improve the radiative flow of heat to the room, later fireplace builders tilted the back wall forward and canted the side walls outward. This practice accorded more or less with Count Rumford's notions about fireplace design. Tilting the rear wall forward near the top not only gained some radiative surface but also narrowed the throat and brought it forward, leaving room for the smoke shelf at the rear. The higher velocity in the throat tends to stabilize flow above it and makes the smoke shelf function better. With the sides of the fireplace at right angles to the back, they simply exchange radiation with each other to no effect. When each is canted outward, the room receives the benefit of radiation from their hot surfaces.

Count Rumford's Fireplace

Heat preoccupied Benjamin Thompson at the practical as well as fundamental level. Shown here from his own papers is his plan for one of the 200 fireplaces he is said to have rebuilt to his more efficient design in the town houses of the rich and powerful in London. This was during his first years in exile from his native land, before his service in the court of the Elector of Bavaria made him Count Rumford. The fireplace of the period was, typically, more than half as deep as it was wide and it vented into a full-throated chimney, as shown in Figs. 4, 3, and 6. With the fire built securely in the rear of the fireplace, the range of its radiation, the only

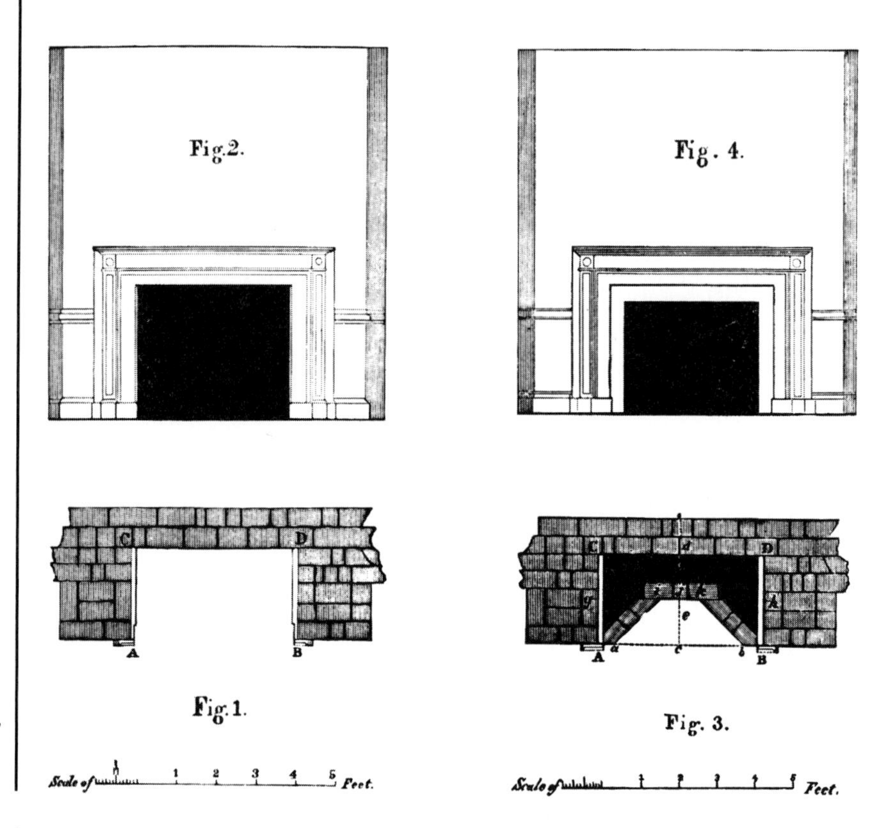

Rumford improved on earlier fireplaces (an example is shown in Figs. 2, 1, and 5) by adding a smoke shelf and canting the side walls of his fireplace (Figs. 4, 3, and 6).

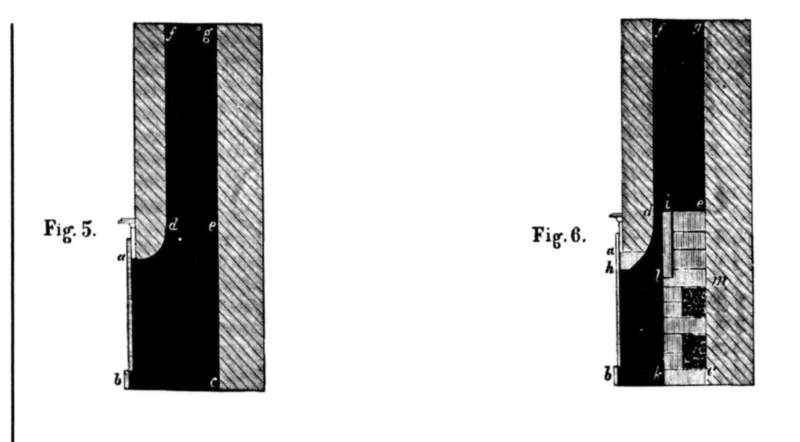

mode in which it contributed to the heating of the room, was narrowed. Fluctuations in air pressure due to winds above the chimney out of doors could roll billows of smoke into the room. Thompson brought the back of the fireplace halfway forward, narrowed it, and tied it to the front of the fireplace with two diagonal walls of brick, as he shows in plan and cross section at the right. The effect of this arrangement was to bring the fire forward, increasing the sweep of its radiation, and to reinforce the radiation with reradiation of its heat from the diagonal walls. The forward placement of the back wall of the fireplace had the additional effect of narrowing the throat of the chimney and providing a smoke shelf at its bottom. Against this obstacle, and against the draft of the fire accelerating through the narrow and then expanding throat, a downdraft would be rolled back and carried up the chimney. From his success in London society, Sir Benjamin went on to find his fortune on the Continent, taking with him a British knighthood.

I have four fireplaces in my old colonial farmhouse in Maryland that reflect these developments in design. The one in the kitchen was built for space heating and is shallow so that the fire is thrust forward and the back wall radiates its warmth into the room. (The original, fifth, cooking fireplace was in the basement and has since been walled in to provide a flue for the central heating plant.) The fireplace in the living room is larger and deeper and in fact is not as effective a space heater as the smaller kitchen fireplace. Neither fireplace is fully symmetrical, probably because the bricks are uneven in size. Their dimensions, shown in the figure on page 46, may be compared with the standards recommended by the U.S. Department of Agriculture, given in the

Two old fireplaces of different size were built to the same proportions. Forward tilting of the back wall improves radiation from the fireplace into the room and creates a "smoke shelf" in the throat of the chimney that deflects downdrafts (see the figure on page 43). The older one was probably rebuilt to add the damper.

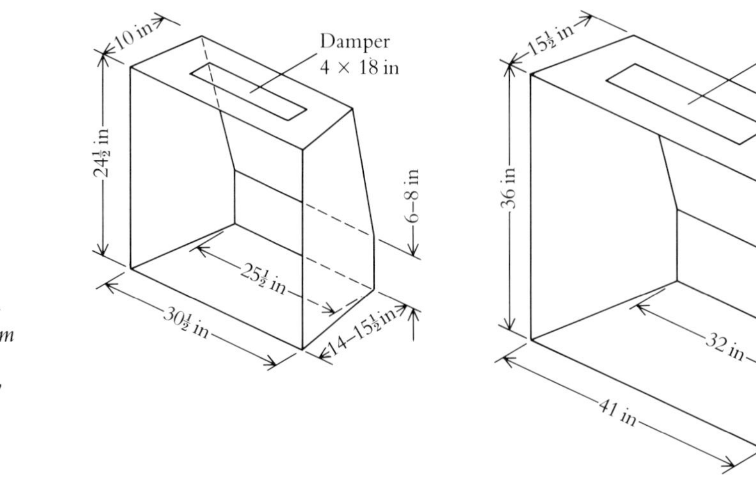

Kitchen fireplace, ca. 1840

Living room fireplace, ca. 1780

table below. Comparison suggests that the vertical, lower section of the back wall in my fireplaces is much too low. Current practice is to have the back wall rise vertically at least 14 inches in fireplaces of this size. Nonetheless, the two old fireplaces function beautifully and I am reluctant to tamper with success. The old house has brick flues but no flue liners, a situation rendered relatively safe by the massive size of the surrounding chimney. In modern times these would have ceramic-tile flue liners of the first and second sizes shown in the USDA table.

Fireplace Dimensions Recommended by the U.S. Department of Agriculture

Fireplace opening proper			Interior chimney dimensions (size of flue liner)	
Width (inches)	Height (inches)	Depth (inches)	Rectangular (inches)	Round (inches diameter)
24	24	16–18	8½ × 13	10
30	28–30	16–18	8½ × 13	10
42	28–32	16–18	13 × 13	12
60	36	18–20	13 × 18	15
72	40	22–28	18 × 18	18

Source: USDA Farmer's Bulletin No. 1889.

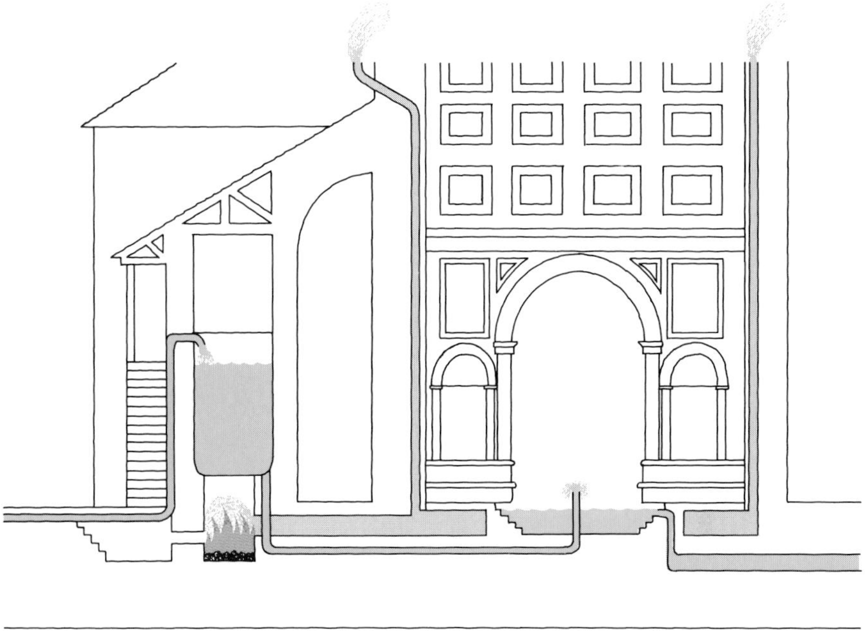

The Roman bath is recalled here as a model central-heating system. The heat of the fire went directly to heat the water, which was piped first to the hottest bath chamber, the caldarium; thence to a warm pool, the tepidarium; and last to a cold pool, the frigidarium (only the hot pool is shown). What otherwise might have been the waste heat of the hot gases from the fire was conveyed under the floor of the caldarium by the hypocaust and to its walls by the flues, through which the smoke was vented.

Houses built around central chimneys were the first to be able to boast of central heating. Over the course of the cold season the heat absorbed into the large thermal mass of the masonry, brick, and stone of the chimney contributed its own warmth to the house. By a similar stratagem the walls opposite fireplaces in medieval castles were built thicker to soak up and reradiate heat. In the great houses of post-Renaissance Europe, big masonry stoves reaching from floor to ceiling and often wonderfully adorned with ceramic tile filled the corner or stood out from the wall of the room. Inside, the thermal mass was tunneled with convoluted paths for the flue so that large masses of stone were heated by the combustion gases. Such installations extracted heat efficiently from the fire and radiated it to the surroundings. They warmed up slowly, however, and the fire inside had to be kept going. (The trick of increasing the area for heat transfer by twisting the flue pipe back and forth can also be used in metal stoves to good effect. In industrial furnaces this principle is taken to its ultimate limit in the design of heat exchangers, as shown in the next chapter.) Thermal masses are incorporated in the design of new homes in newly fuel-conscious America, especially in passive solar schemes.

The fireplace, like the candle, serves a more ceremonial than practical function in our time. Even when people had to depend on it for warmth they

Exposed here is the hypocaust, open spaces under the floor, of a Roman bath, with stumps of the columns that supported the stone floor and with one section of the floor still in place on its columns. Hot gases from the fire that heated the water were circulated through the hypocaust to warm the floor.

appreciated it more for its aesthetic amenities. Benjamin Franklin observed, ". . . A Man is scorch'd before, while he's froze behind." The relative role of radiation and convection in transferring heat from the fire on the hearth to the room has only recently, however, been reduced to numbers.

It is now clear that the flow of air by convection is almost entirely from the room into the fireplace and up the chimney. Heating of the room is thus almost entirely by radiation. Radiation only produces heat when there is a mechanism for absorption. Air is little heated by radiation from a wood fire. The radiation strikes various objects in the room, is absorbed, and the temperature of the objects rises. The air will remain cooler than the objects in the room. If you warm yourself in front of a fire, only the side facing the fire benefits; as Franklin observed, the other side remains quite chilly.

Recent studies at Lawrence Berkeley Laboratory of fireplace efficiency show that only 12 to 14 percent of the heat of combustion in a fireplace is delivered to the room; the rest is lost up the chimney. As if this were not bad enough, the pumping action of the fire draws room air up the chimney in amounts 10 to 15 times that needed for combustion. This air is replaced in the dwelling by cold air that infiltrates through cracks and crevices, especially around windows and doors. Infiltration of excess air offsets the heat gained from the fire. The result is a cut in efficiency to about 6 percent: the unadorned fireplace manages to lose about 95 percent of the available heat represented by

Benjamin Franklin portrayed in London in 1767.

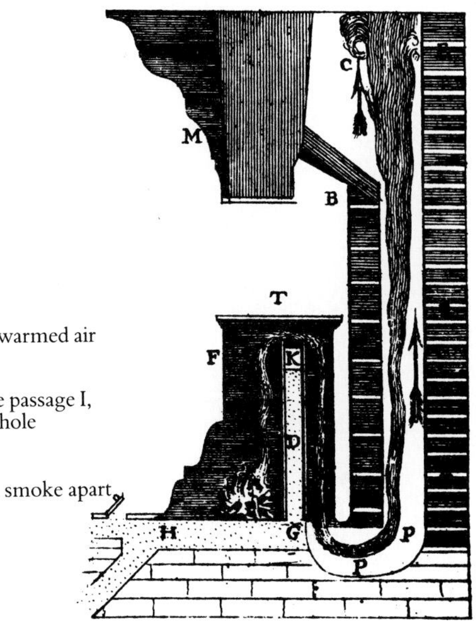

M The mantelpiece or breadth of the chimney

C The funnel

B The false back and closing

E True back of the chimney

T Top of the fireplace

F The front of it

A The place where the fire is made

D The air box

K The hole in the side place, through which the warmed air is discharged out of the air box into the room

H The hollow filled with fresh air, entering at the passage I, and ascending into the air box through the air hole in the bottom plate near

G The partition in the hollow to keep the air and smoke apart

P The passage under the false back and part of the hearth for the smoke

↑ The course of the smoke

The Franklin stove, an early invention of the foremost American natural philosopher, cured many of the defects of the fireplace. Franklin's own explanatory diagram of the separated circulation of the room air and fire smoke in his stove is shown. The stove enclosed the fire, brought it out to the front of the hearth, and increased energy output by giving heat to the room air by conduction and convection from its metal surfaces, from which, in addition, it gave out much heat by radiation. With the doors closed it allowed just enough air into the fire to keep it going and stopped the loss of warm room air by convection up the chimney (see the figure on page 43). Room air was drawn instead through a "hollow" under the fire into an "airbox" around the fire, warmed there, and returned by convection to the room. The name Franklin was later misattached to almost any kind of stove set out in the room.

that pile of wood so carefully stacked out behind the shed in the fall. What is more, these efficiencies were reported by Lawrence Berkeley Laboratory for the relatively moderate outdoor-indoor temperature difference of 20°C, or 36°F. For appreciably larger differences the infiltrating cold air will overwhelm the effect of the fire and there will be a net heat loss from operating the fireplace. This occurs at a temperature difference of approximately 28°C, or 50°F.

Benjamin Franklin was the first to attempt a significant improvement over the fireplace on this side of the Atlantic. He won early fame for the stove he invented in 1740, 60 years before Rumford attacked the problem. Franklin's device was very unlike the stove we now associate with his name. Whereas today a Franklin stove is simply a metal fireplace moved out into the room to allow convection heating by air flow over the external surfaces, Franklin's first stove was far more sophisticated. He described its design and the motivation for it in a pamphlet in 1744: "An Account of the New Invented Pennsylvanian Fire Places, Philadelphia, Printed and Sold by B. Franklin." That document presents a fair analysis of the properties of fire and the defects of the then-extant fireplaces and stoves. Franklin discusses the severe drafts caused by the large open chimneys and the near total loss of heat to the room this brought about. It is clear that he understood very well the phenomena governing the operation of a fireplace. About the only error I can detect is his belief that air would absorb appreciable quantities of radiant energy; that effect is small.

Count Rumford's Oven

With the iron stove coming into use in 18th-century households to supplement and replace the fireplace for heating and cooking, Rumford made a truly original invention. Baking and roasting, in households that could afford such luxury, were done in a chamber built into the brickwork alongside the fireplace with its own flue connecting it to the chimney. When the brick or stone lining of this chamber, with its high heat capacity, had soaked up enough heat from a fire built inside it, the coals and ashes were raked and swept out and the food put in place to be cooked by the reradiated heat. By this sequential procedure the fire was made to convey its heat to the food without the flavor or taint of its smoke. In the iron stove, Rumford saw another way to keep the smoke out of the cooking. He put a chamber into the stove, sealed away from the fire but heated by its combustion gases conducted through surrounding flues. Through the iron, with its high heat conductivity, the heat of the fire was supplied steadily to the foods. Shown at left and below is the Rumford Roaster. Not until the smokeless gas flame and electric heating element came on the scene did the heat source get back inside the oven.

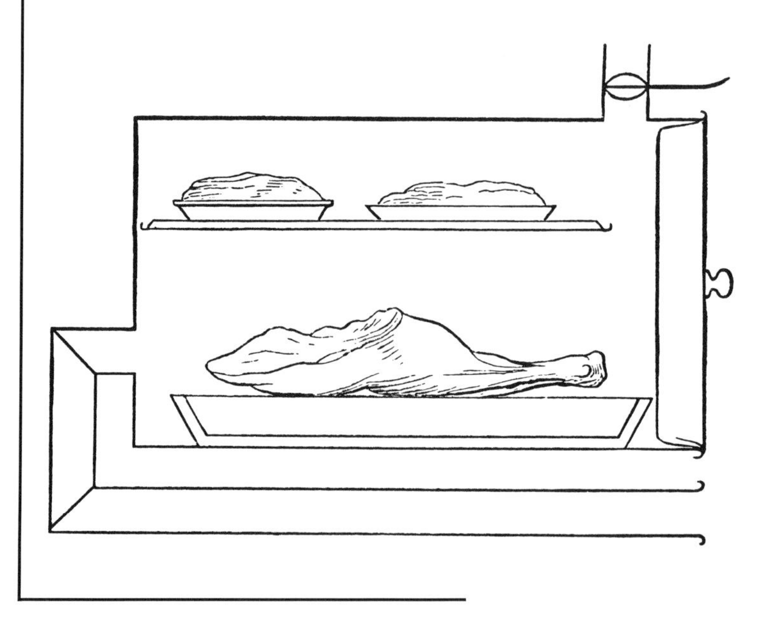

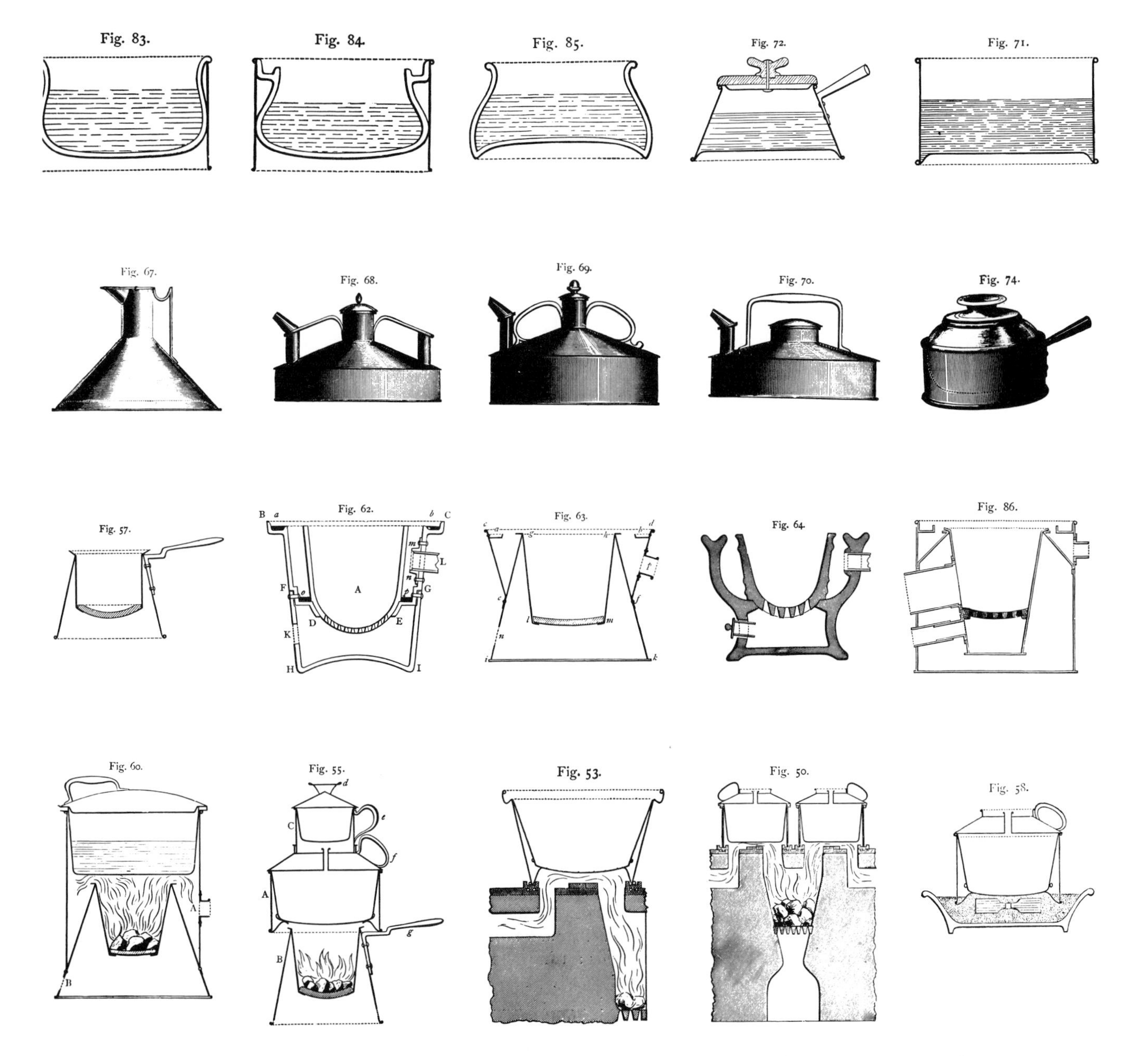

Count Rumford gave active attention to the other principal domestic use of heat: for cooking. His designs for saucepans and kettles (top two rows) maximized uptake of heat with a high ratio of the horizontal to the vertical dimension. In the third row from the top he conjured with various designs for steamers and in the fourth, taking advantage of the new availability of sheet iron, he invented self-contained pot-and-fire units, plus early versions of the modern kitchen range with a small cooking fire under the pot(s).

Franklin designed a metal box, open at one end, fitted with an internal heat exchanger that he termed an air box. The metal box was set in the fireplace opening and the opening itself was bricked up in a false back with a channel provided for combustion products to reach the chimney proper. The stove front was fitted to receive a shutter sliding vertically to close off combustion air when desired. There was also a damper in the rear channel. A separate air stream from the room flowed under the hearth and entered the air box, where it passed upward through convoluted channels cast into iron facing plates and flowed back into the room from openings near the top. The room was heated by convection from the air box, by the flowing of the room air over all the metal surfaces, and by radiation from all the surfaces and from the fire itself.

Franklin was offered a patent by the governor of Pennsylvania. He declined with the famous quotation: "As we enjoy great Advantages from the Invention of others, we should be glad of an Opportunity to serve others by an Invention of ours, and this we should do freely and generously."

In domestic American life, the original and then the modified Franklin stoves were steps on the way to the standard potbellied stove. That stove appeared in about 1836 and did much of the indoor heating in the United States during the 19th century.

The recent fuel consciousness has brought on the market a large variety of stoves using the so-called air-tight principle. They are, of course, not air-tight, but the air flow is controlled so that not much more air than is needed for combustion is admitted. This change cures the loss in efficiency from excess convection of room air into the chimney. J. W. Shelton of Williams College has measured the efficiencies of certain "air-tight" stoves at 50 to 60 percent. He points out the significance of the radiation and convection from the stovepipe; his figures indicate that up to one-fourth of the total heat comes from a six-foot pipe.

Franklin's original design and his attachment to a visible, cheering fire persist today in efforts to improve the fireplace itself. A modest new industry offers a variety of solutions. Convection channels are provided that draw room air in from the floor level, warm it in the circulation around the fire, and return it to the room above, just as in the original Franklin stove. A blower may be installed in the system to improve the throughput.

With or without a convection system, a major barrier to the loss of warm room air can perhaps be more readily installed. A glass screen closing the fireplace opening but permitting a sufficient flow of air to sustain the fire serves this purpose. The poor transparency of ordinary glass to infrared radiation (ordinary glass becomes opaque at a thickness somewhere between 2½ and three micrometers) can be cured at extra expense by the use of special glass or natural quartz windows that transmit more infrared radiation and offer more resistance to thermal damage.

The potbellied stove was standard equipment of U.S. households throughout the 19th century. A good deal more efficient than a fireplace, it heated the room air by conduction and convection over its hot surfaces, including the length of its stovepipe. It also heated by radiation from its surfaces, but it hid the flames.

The closed door front on the fireplace or Franklin stove has a great advantage over an open fireplace after the fire has died down. The throttled air flow holds down losses up the chimney during the night when the ashes are still hot and combustion gases are still emanating from the residues. (The damper in the chimney should never be closed until the ashes are cold to the touch: toxic gases, such as carbon monoxide, will otherwise back up into the dwelling.) Without a glass door, an open damper in the wintertime permits large convective losses even when the ashes are cold. This is from the so-called stack effect, arising from the difference in air density inside and outside the

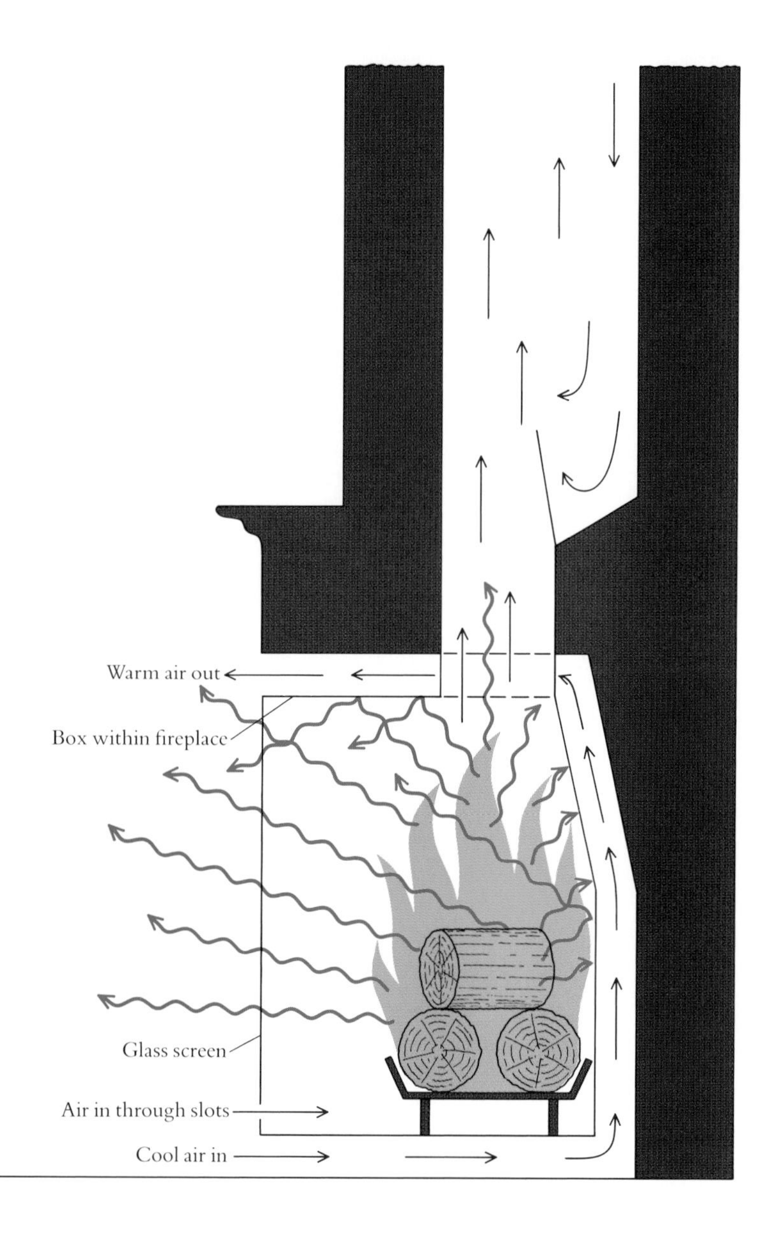

Fireplace efficiency can be improved by a combination of devices. A panel of quartz or glass of suitable transparency (see the figure on page 55) will permit radiation from the fire to reach the room and will keep warm room air out of the chimney convection stream (an opening at the bottom admits enough air to keep the fire going). A metal insert, modeled on the Franklin stove (see the figure on page 49), can be fitted into the fireplace, permitting the room air to be drawn in under the hearth, circulated around the firebox, and returned by convection to the room.

house; at a difference in temperature of 40°C the stack pressure difference comes to about 0.1 inch of water pressure for every 10 stories of building height. In a skyscraper this difference in pressure may create a veritable gale; in a two-story chimney it creates a breeze, gentle but strong enough to waft away a great many energy dollars each day. The answer is to keep the damper closed when the fire is dead and to use transparent doors fitted with air slots when it is alive.

In the 20th century—at least until the recently rekindled interest in room heaters—the whole problem has been relegated to the basement. We gave up the age-old comfort of sitting in the flickering light of the wood fire in ex-

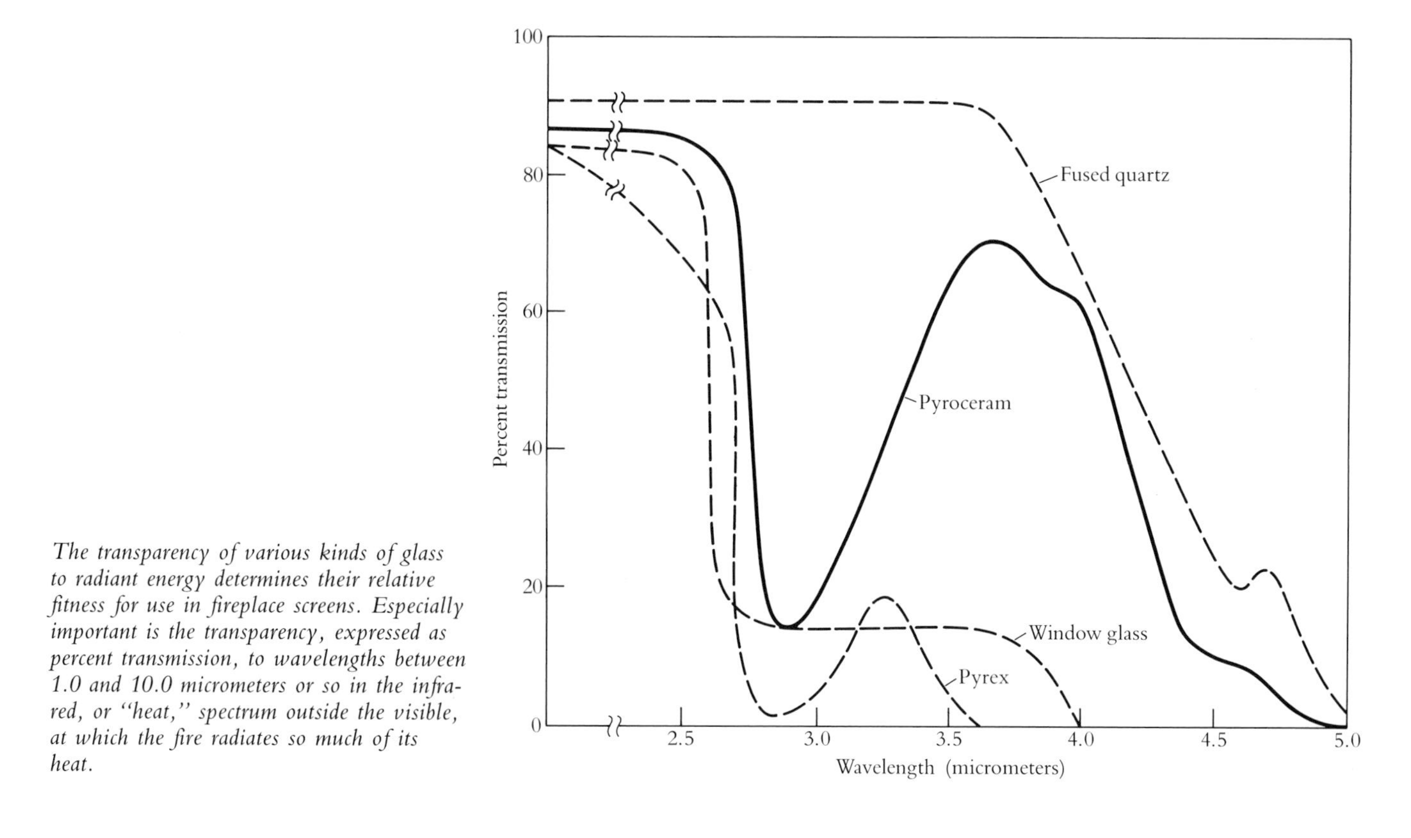

The transparency of various kinds of glass to radiant energy determines their relative fitness for use in fireplace screens. Especially important is the transparency, expressed as percent transmission, to wavelengths between 1.0 and 10.0 micrometers or so in the infrared, or "heat," spectrum outside the visible, at which the fire radiates so much of its heat.

change for high-efficiency, uniform heating from central furnaces. The modern furnace has a carefully controlled feed of combustion air, its heat-transfer surface optimized to bring out as much heat from the flue gases as is economically practical, automatic dampers to control backdrafts, and thermostatic switches to maintain nearly constant temperature. The fire gases are totally separated from the circulating air of the home and in many cases hot water or steam is the heat-transfer fluid, not air. The whole is very efficient—on the order of 75 percent and, in some cases, over 90 percent—but it is not very romantic.

A basic oxygen furnace boils furiously as a lance feeds oxygen to its molten iron charge. Oxidation of carbon in its iron supplies heat.

4 *Furnaces and Combustion Chambers*

The steady flame of the candle and the flickering flames on the hearth shed their light and warmth on our lives at home. We have regarded them closely as the tame, domestic incarnations of a genie that, running out of hand, can destroy our homes and take our lives. Over the past three centuries of industrial revolution, that same genie, contained and harnessed in furnaces and combustion chambers, has changed the map of the world and revised the terms and condition of existence of one-third of humanity. In this chapter we consider how the heat of fire is put to work. We shall make a friendly acquaintance here with its enormous power.

A furnace is a container employed to apply the heat of fire in order to effect a physical or chemical transformation in some material or to transform heat with the greatest efficiency into mechanical work. Furnaces variously fire ceramics; melt sand, soda ash, and some trace ingredients into glass; reduce or smelt metals out of ores; raise high-pressure steam to drive turbines that spin electric generators; and, more humbly, fill household radiators with hot water or steam. For whatever purpose, the principles of furnace operation are everywhere the same. In adaptation to its particular application, each furnace acquires its own personality.

As mentioned in our consideration of the candle, the transfer of heat from the fire to its point of useful function proceeds the faster, the greater the temperature of the fire and so the greater the temperature difference between the source of the heat and its sink. That difference is the thermal driving force. Conduction of heat through a solid proceeds as a function of the first power of the temperature difference across the solid. This is Fourier's law, in the elaboration of which Jean Baptiste Joseph Fourier wrote an algorithm indispensable for the analysis of waves of any kind (*see the box on page 58*). Transfer of heat by radiation between two bodies proceeds as a function of the difference between the temperatures of the two bodies, each raised to the fourth power. Clearly, the first principle of furnace operation must be to seek the highest possible—that is, manageable—flame temperature.

The highest-possible temperature is that reached in the theoretical situation where no heat is allowed to leave the flame; this is the so-called adiabatic

Fourier

Jean Baptiste Joseph Fourier remains today a model of the intellectual civil servant who has served France so well. An offspring of landed gentry at Auxerre, south of Paris, and orphaned at the age of eight, he was educated at the military school in Auxerre and, at 16, became a teacher of mathematics there. He survived the Revolution to join the founding faculty of the great École Normale. In 1798, serving on the staff of Napoleon, he was a member of the expedition to Egypt. As governor of Lower Egypt, he mobilized local resources to support the occupying French army when the British cut its supply lines to home. For his service there and later in administrative posts in France, he was ennobled in 1808. Settling in Paris in 1816 as professor of the Polytechnic Institute, he succeeded Pierre Laplace as president of the governing council of that distinguished center of learning and died in that post.

In service at Grenoble as prefect of Isère he undertook his mathematical investigation of the conductance of heat in solids, a fundamental investigation stimulated by rising interest in heat engines, the steam engine to begin with. In his *Théorie Analytique de la Chaleur,* he not only resolved that question with elegance but made a basic contribution to mathematics.

He found that the rate of conduction of heat through a body is a function of the difference in temperature between the hot side and the cold, with the conductivity of the body a constant. Fourier first determined that the flow of heat through a body whose boundaries are held at two different but constant temperatures obeys a very simple rule. This is now known as Fourier's law of heat conduction:

$$\text{rate of heat flow} = -\text{constant}\left(\frac{T_2 - T_1}{X_2 - X_1}\right)$$

or, in terms of differential calculus:

$$\frac{dq}{dt} = -k\frac{dT}{dX}$$

where q is heat, t is time, T is temperature, X is the position coordinate, and k is the thermal conductivity of the body. In 1823 Fourier published a complete treatise based on this equation, elaborating it with a mathematical analysis describing many different kinds of heat-conduction problems. In the treatise, he provided a general differential equation that related temperature, time, and spatial coordinates. By manipulating the solutions to this equation, he found ones that are infinite series of trigonometric functions (sines and cosines). From this came the technique of constructing any arbitrary wave form by using a suitable series of such functions. With respect to any function $f(x)$, in Fourier's words: "We can always develop this function in a series that contains only sines or cosines or the sines and cosines of multiple arcs."

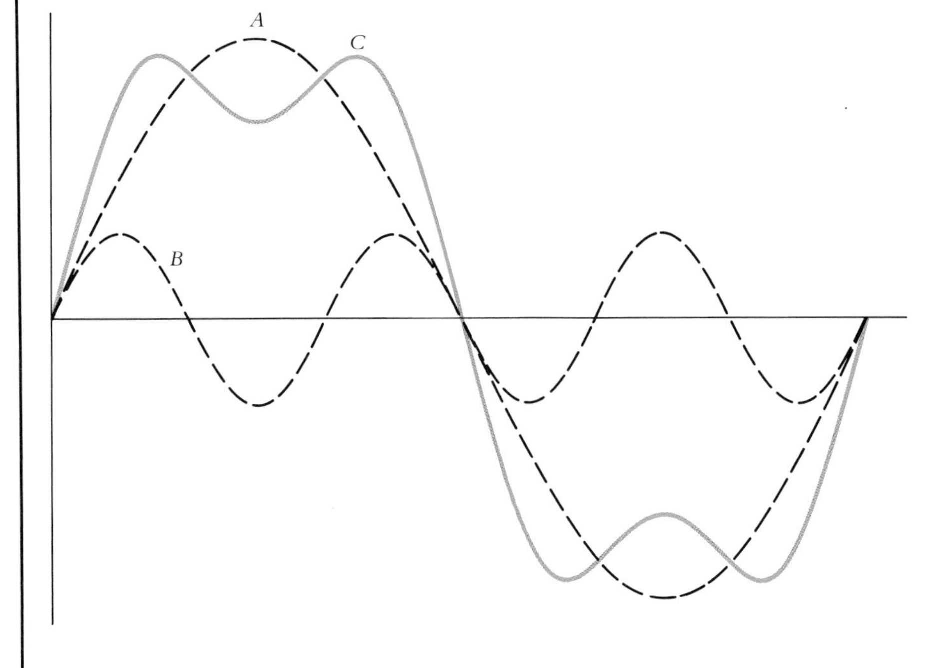

One can also do the reverse: render an arbitrary wave form in terms of its constituent trigonometric functions. This is called Fourier analysis, a technique much used in electronic signal processing. In the illustration above, for example, Fourier analysis will resolve a compound wave like *C* into its two component sine waves, *A* and *B*.

temperature. It may be calculated from the heat of combustion of the fuel and the heat capacities of its combustion products. The equation reads simply enough:

$$\Delta H = n_1c_1(T_f - T_0) + n_2c_2(T_f - T_0) + n_3c_3(T_f - T_0) + \cdots$$

where ΔH is the heat released, n represents in turn each of the combustion products, c is the heat capacity of each n, T_f is the flame temperature, and T_0 is the initial temperature of the fuel and air.

The reality to be calculated is more complicated. In high-temperature flames the combustion products (and their intermediates) may be molecular fragments whose quantities must be computed from the high-temperature equilibria of the reactions producing the fragments from whole molecules, if known. The production of some intermediates, such as the breakup of water into hydrogen and hydroxyl radicals, is endothermic. The reaction steals energy and lowers the ultimate flame temperature.

The results of such calculations for a sampling of hydrocarbon fuels are shown in the table below. It can be seen that any choice of fuel from that family does not make for more than a 15 percent difference in temperature. Toward reaching the maximum flame temperature, the same table suggests another approach. The temperature-depressing effect of nitrogen, present in the air at four times the abundance of oxygen, is dramatic. The flame temperatures in pure oxygen go 1,000 degrees higher. Those whose interest in science survived into high school will recall the laboratory experiment in which a glowing piece of punk thrust into a test tube of oxygen burst into a white-hot

Some Calculated Adiabatic Flame Temperatures*

	Absolute temperature (K)†
Acetylene in air	2,600
Acetylene in oxygen	3,410
Methane in air	2,232
Methane in oxygen	3,053
Hydrogen in air	2,400
Hydrogen in oxygen	3,080
Heptane in air	2,290
Heptane in oxygen	3,100

From Glassman.

*Assumes fuel and oxidizer begin at room temperature.

†298 kelvins (K) = 25°C.

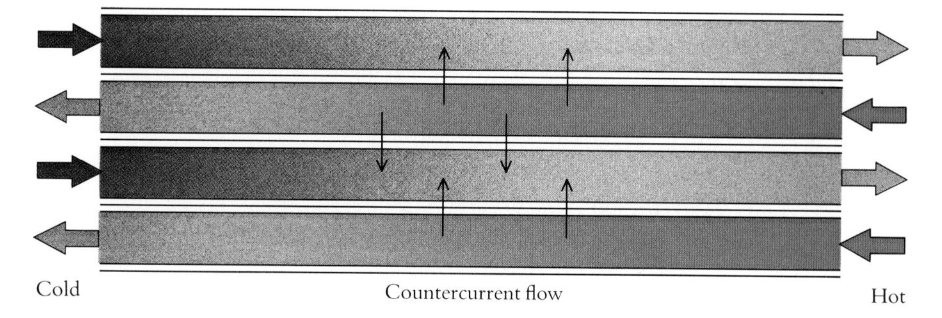

Countercurrent flow causes counterflowing streams to exchange their energy with high efficiency. Heat in the flow (of a gas or fluid) from the right is conducted into the counterflowing stream driven by the temperature gradient across the material that separates the two streams (vertical arrows). In a perfect system each flow would exit from the system near the entrance temperature of the counterflow. What works for heat in a boiler works for the excretion of the products of protein metabolism in the kidneys.

flame. That is why the airline safety drill asks people in the back rows to tamp out their "smoking materials" the moment they see the oxygen masks drop from their overhead compartments.

In air the maximum flame temperature is reached at or very close to the air-to-fuel ratio that supplies exactly enough oxygen to consume the fuel. The temperature falls off with too much air as steeply as it climbs from with too little air. The flame is cooled by the gases, including excess oxygen, that do not go into the combustion reactions. A strategy for increasing flame temperature for a given fuel at once suggests itself: precisely match the supply of oxygen (or another oxidizer) to the fuel supply and, at this setting, reduce the amount of nitrogen and other diluents.

Toward still higher temperatures, another inspection of the equation on page 60 suggests a simple strategy. The usual calculation sets T_0, the initial temperature of the fuel, at room temperature, as specified in the table on page 60. An increase in that temperature results in an increase in flame temperature. That can be easily accomplished by preheating the incoming air and fuel. Heat for this purpose is available in the exhaust of the furnace. Most large furnace installations are equipped to do just such preheating. The fuel and air enter the furnace through a countercurrent heat exchanger. When such a system is working right, each of the counterflowing streams exits from the system close to the other's entering temperature.

With this and other methods it is possible to produce flames that are too hot to handle. The push toward higher temperatures has driven the technologies that supply the materials of furnace construction. In a furnace the heat reaches the charge or the heat-transfer surface of the boiler not only by direct radiation from the flame itself but by reradiation, as in a hall of mirrors, from the walls and ceiling of the furnace. The continuous "black-body" spectrum of the reradiation facilitates the radiative mode of heat transfer. There is radiation as well from the combustion gases downstream, but a good deal more of the heat transfer is by convection, which brings the gases into direct contact with the charge or heat-transfer surfaces. Most furnaces run hot enough to require refractory linings that can stand higher temperatures than the temperature required for their charge. The importance of the reradiation from those surfaces is reflected in the care with which their geometry is arranged to direct and focus it.

The heat passes into the charge or the heat-transfer surface of the boiler by conduction (and in a molten, boiling charge by convection as well) and is lost beyond the refractory lining by conduction through the walls of the furnace and reradiation from their outer surfaces. Insulation of these structures is another active technology.

The overall efficiency, E, of furnaces is summed roughly in the following equation:

$$E = 1 - \left(\frac{T_s - T_a}{T_f - T_a}\right)$$

where T_s is the final off-gas (stack) temperature, T_f is the flame temperature, and T_a is the ambient temperature.

When the hot gases are used to reheat the incoming fuel and air to make low-pressure steam and to heat air for space heating, the stack temperature is reduced to a minimum. Furnaces frequently operate, by this reckoning, at an efficiency of well above 90 percent. Of course, the ultimate measure is how reliably they perform the task for which they are designed.

The first task of the furnace, at the dawn of settled agriculture, was the firing of earthen pots for the storage of grain, oil, and drink. A temperature on the order of 600 degrees is needed to drive the water of crystallization from the clay. This cannot be done quickly, lest pockets of steam build up in the material and blow it apart. The setting of the time-temperature schedules for firing clay remains the key consideration and very much an art today. Models of the heat-transfer processes to predict the firing cycle for a new line of products invariably undergo pilot-plant tests before being turned over to the plant operators, who do their own empirical tuning of the final cycle.

Glass making is another process in which the flame and the furnace atmosphere supply heat for the rearrangement of the chemical bonds of the simple ingredients that make this improbable end product. The reactions occur with the melting of the ingredients together in temperatures that reach 1,500°C. Except in the making of fancy crystal or highly technical products, glass making goes continuously, not by the batch. Clean sand premixed with one or another source of metal oxides is fed in at one end and molten glass ready for forming flows from the other at a rate of 700 tons a day, from modern installations. The long, shallow glass furnace is aptly called a tank. Flames of gas or oil premixed with air heated in firebrick-lined regenerators flood across the melt alternately from either side under the incandescent fire-

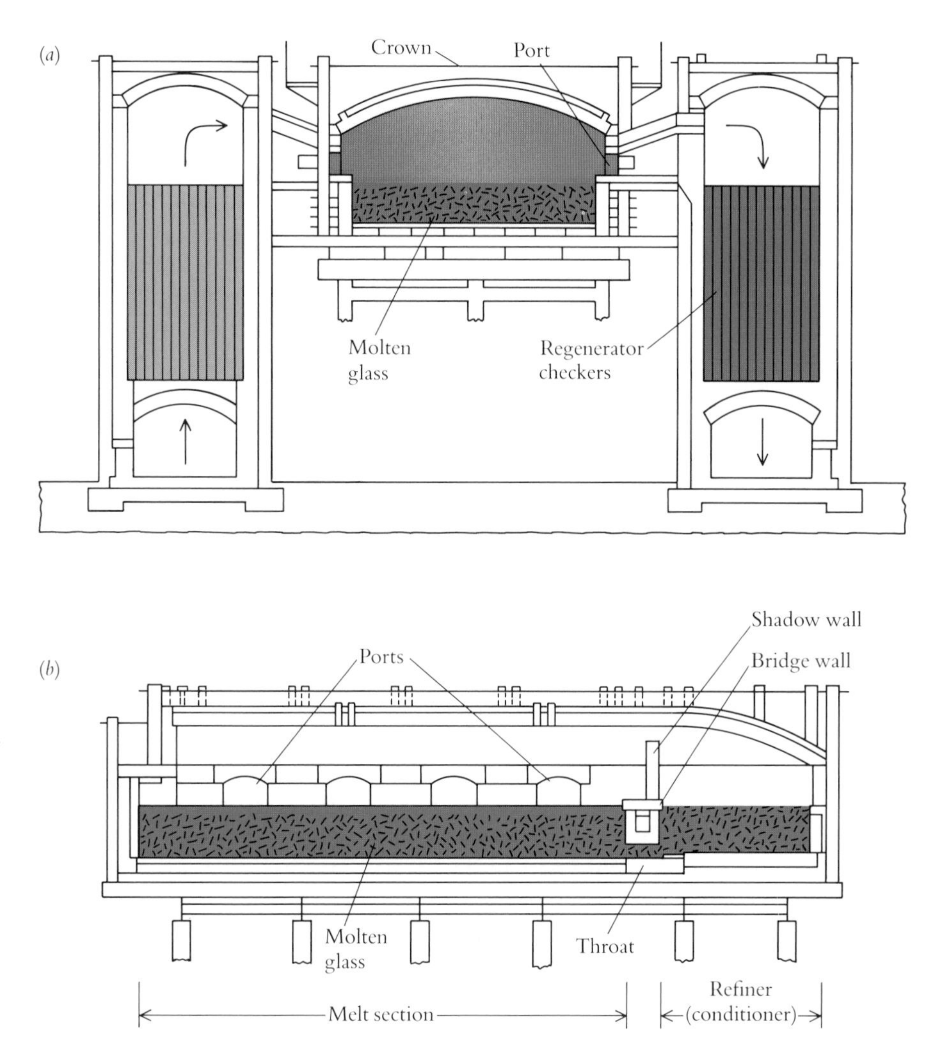

A glass furnace, shown here in transverse (top) and longitudinal (below) cross section, operates continuously. Raw materials go into the furnace at left in the longitudinal cross section. Flames (natural gas or petroleum) and hot combustion gases sweep across the melting and molten charge from ports along the length of the melt section; the product flows from the furnace for further processing at right. The transverse section shows regenerators that conserve heat in the operation of the furnace. Combustion gases are exhausted and drawn alternately into the regenerator on one side and then on the other. The heat left behind by these gases is invested in the new air drawn in through the hot regenerator. This heat goes to increase the temperature of the next cycle of flames that sweep across the furnace. The arch of the ceiling focuses the reradiation of heat from its firebrick on the melt.

brick lining the arched ceiling. From the glass-making reactions the stacks carry off nothing more noxious than carbon dioxide.

Copper smelting, the intermediary step that isolates the copper as a sulfide from its ores, is conducted continuously in a similar furnace. The charge enters the furnace under the flames at the upper end. Melting into two layers, slag above and the copper "matte" below, it gathers in a large pool at the lower end. The multiple reflections of radiant heat at about 1,000°C from the cherry-red sides and arched ceiling gives this reverberatory furnace its name. The furnace gases are particularly noxious because of the sulfur and other oxides swept up in the atmosphere. Regenerators are not used, but there is some recapture of downstream heat, in waste boilers and other devices.

In the making of steel, the furnace fire actively engages in the chemistry that wins the iron from its oxide and adjusts the percentage of carbon in the finished metal. The reactions are exothermic and themselves supply much of the heat. Iron was produced for nearly 3,000 years by packing ore and charcoal into small furnaces. The early iron masters could not get these furnaces up to the melting point of iron, but they could get them to the temperature, about 500°C, at which the oxygen would move preferentially from the iron to combine with carbon from charcoal and go off with it in the flame. The iron so made had a very low carbon content and could be easily hammered into shape. Such wrought iron could also be too easily bent out of shape.

In Europe, beginning in the 14th century, leather bellows operated by foot or by waterwheel, and then water-powered reciprocal air pumps, supplied an air blast to make these primitive furnaces hot enough, above 1,000°C, to melt iron. Cast iron from these furnaces, with a carbon content of 5 percent and more, was too brittle to be wrought. Although armorers had for centuries known how to get just the right amount of carbon (about 1 percent) into Damascus and samurai blades, the techniques were kept secret. The age of steel had to await the Bessemer converter and the open-hearth furnaces of the mid-19th century.

The copper converter separates sulfur from copper in the copper matte from a reverberatory furnace in a cycle of exothermic reactions. Copper retained in the slag from this furnace is recycled in the charge of the reverberatory furnace.

To supply the appetite of these steel-making furnaces, the blast furnace assumed its modern form. The flow of material and heat in this furnace, as in its ancestors, is vertical. The charge of ore, limestone, and coke goes in at the top. There are no burners; all the fuel is in the coke. Air from one of the three tall stoves lined up beside the furnace enters it through ports at the bottom, a hot air blast at about 600°C. Oxygen in the air blast oxidizes some of the coke in the exothermic production of still hotter carbon dioxide. In an endothermic reaction the carbon dioxide reacts with unspent coke to make carbon monoxide. This gas reacts exothermically, at more than 1,000°C, with the iron oxide to carry off the oxygen as carbon dioxide and free the iron. The hot gases from the top of the furnace heat the brickwork checkers in the next stove to sustain the blast of hot air at the bottom of the furnace and keep the cycle going:

$$\underset{\text{coke}}{C} + \underset{\text{oxygen}}{O_2} \rightarrow \underset{\text{carbon dioxide}}{CO_2} \quad \text{exothermic}$$

$$CO_2 + C \rightarrow \underset{\text{carbon monoxide}}{2CO} \quad \text{endothermic}$$

$$3CO + \underset{\text{iron oxide}}{Fe_2O_3} \rightarrow \underset{\text{iron}}{2Fe} + 3CO_2 \quad \text{exothermic}$$

The carbon dissolved in the cast iron from the blast furnace fuels an exothermic reaction that adjusts its presence to just the right percentage in the

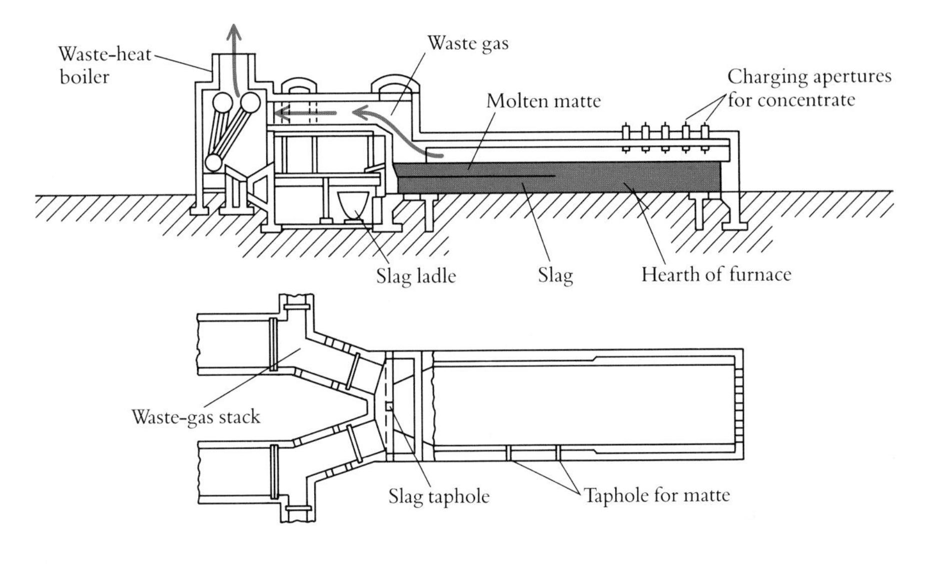

A reverberatory copper-refining furnace captures copper sulfide from its prepared ore in a molten mass. Such a furnace, shown here in longitudinal cross section and plan, is charged with ore and flux materials, which help to remove impurities, at right. From fuel burners at that end of the furnace flames sweep down the furnace over the charge and into the flues at left. The geometry of the furnace lining focuses its radiation on the melt. The melting charge separates into two layers, with the slag above and copper matte below. The matte is drawn off for the release of the copper from its sulfide in the converter. Circulation of the combustion gases in this furnace is all one way, in contrast with the glass and blast furnaces (see the figures on pages 63 and 66); although heat exchangers capture some boiler heat from the flue gases, fuel consumption is high.

steel in either the Bessemer or open-hearth furnaces. These romantic installations, the Bessemer belching its ruddy flame and the open hearths with their tall stacks drawn up in a straight line, have yielded to the basic oxygen furnace. With the bulk isolation of liquid oxygen from the air, an accomplishment of the years following World War II, this furnace has speedily come to make more than half the world's steel. In the big, potbellied vertical converter lined with firebrick, the charge of molten pig iron brings with it both heat and carbon fuel. A lance delivers pure oxygen gas, and the strongly exothermic reaction does the rest.

Some of the principal reaction vessels of the modern chemical industry have the harsh characteristics of the furnace, and the technology owes much to experience with metallurgical and ceramic furnaces. A catalytic cracker for re-forming the molecules of raw petroleum into gasoline, standing 65 meters tall and 15 meters in diameter, is as imposing as a blast furnace. Inside, the gaseous petroleum rushes in a kind of cool flame at 500°C, close to its ignition temperature. On the vast surface of catalyst, exposed as grains of aluminum silicate suspended in the upward flow, the chemical bonds of the petroleum molecules are rearranged. The heat is supplied externally by burning coke to heat a regenerator that restores the catalyst and returns it to the reactor at 600 to 700°C.

A very different sort of furnace is the steam boiler. Its sole purpose is to "raise steam"; that is, to vaporize water into steam at high pressure and temperature, economically and quickly, and thereby to move as much as possible of the thermal energy of the fire to the world outside where it can do useful

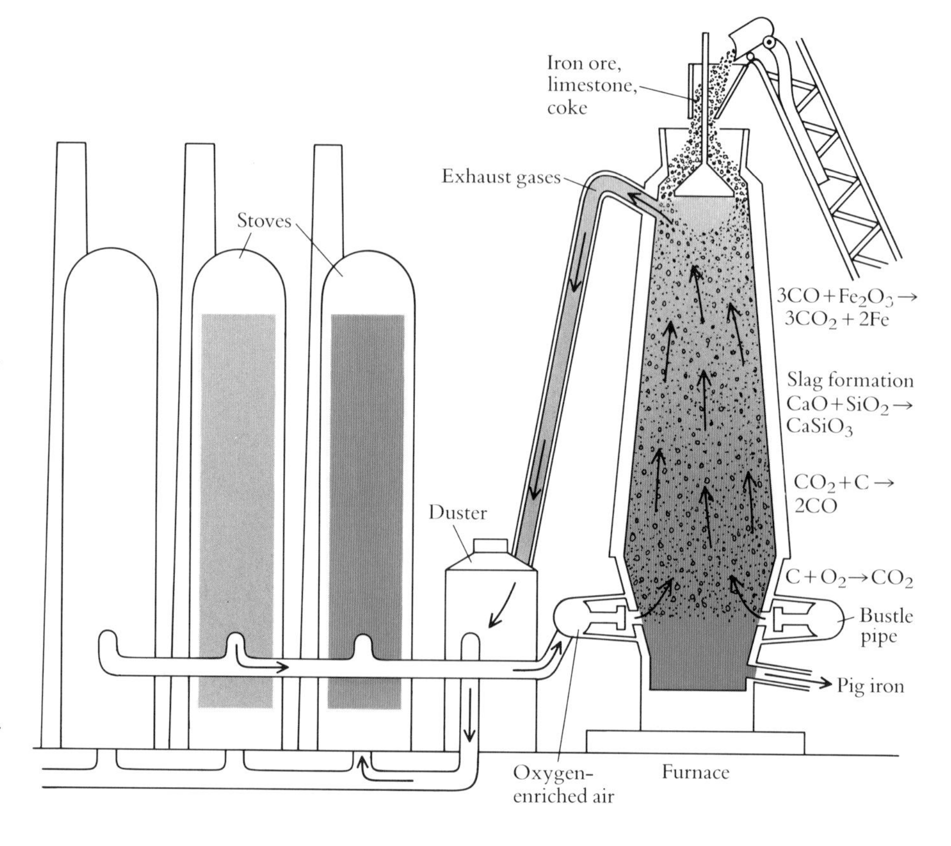

A blast furnace engages the oxygen from the oxide of iron in the ore, securing the combustion of the carbon in the fuel and generating the heat that frees the iron from its oxide. Iron ore, coke, and fluxes are fed into the top of the furnace. A continuous blast of hot air fed into bottom of the furnace from stoves (left) starts the cycle of reactions. Carbon from the coke is oxidized in exothermic combustion to CO_2 in the first zone; in the hotter zone above, carbon dioxide shares its oxygen in an endothermic reaction with the more abundant carbon in the coke to generate carbon monoxide. Carbon monoxide nearer the top of the furnace picks up oxygen from iron oxide in exothermic reactions, issuing carbon dioxide in exhaust gases, and molten iron starts to trickle toward the bottom of the furnace. Exhaust gases, cleaned in the duster, leave their heat for the next incoming stream of air in one of the three stoves that stand alongside the furnace.

work. Generally, the more compact the plant, the more efficient it is. Steam engineers have given their best creative effort to improving the contact between the fire gases and the surfaces of the vessels and tubing on the other side of which the water turns to steam. They either run the fire gases into ducts passing through the water reservoir (fire-tube boilers) or the water into tubes through the firebox and flues (water-tube boilers). To achieve the intimate contact of gas and metal, they take measures to secure high velocity and turbulence in the flow of the fire gases in order to reduce to a minimum the thickness of the stagnant and insulating boundary layer of gas next to the surface. Many passes of tubes placed in the torrent of gases on their way to the stack in a water-tube boiler help to secure both extensive surface exposure and passageways that set up turbulence.

Heat transfer is secured as well from the 25 or so percent of the flame energy that goes out as radiation. Boiler engineers favor oil and coal over natural gas because these fuels yield a larger quantity of incandescent soot particles in their flames. As in the yellow tongue of the candle flame, the radiation has a continuous spectrum, of such power as to swamp the band spectra from the atoms and molecules in the gases.

The movement of energy by radiation from one surface to another, as has been mentioned, goes as the difference between the temperatures of the two surfaces, each raised to the fourth power. That principle is quantified by its more precise statement in the equation:

$$\Delta E = \epsilon\sigma(T_f^4 - T_t^4)$$

where ΔE is the change in energy, ϵ is the emissivity of the surface, σ is a constant (Stefan-Boltzmann), T_f is the flame temperature, and T_t is the tube temperature.

Emissivity, a property of the radiant body, is rated on a scale from 0 to 1. In a flame it is a function of the number of radiating particles. In the nonluminous gas flame it may be as low as 0.1. In coal and oil flames it approaches the black-body ideal of 1.0. At the receiving end in a boiler the emissivity of the tubes also has a high value, approaching 1.0 if there is some soot on them (a virtue that is offset if there is so much soot as to insulate them).

Boiler engineers run the hottest furnaces they can. In pushing for hotter flames, they must not allow the furnace linings to overheat. They handle this by careful choice of refractories and by geometry. The tube surfaces remain, of course, close to the temperature of the water or steam on the other side, and it is fast conduction of heat through the metal into the water and steam that the engineer seeks. In the firebox the flame temperature may reach 1,600°C; the off-gas temperature may be brought as low as 120 degrees. Calculated by the equation on page 62, the overall efficiency of the furnace exceeds 90 percent. To the world outside, the furnace delivers steam at temperatures exceeding 500°C and pressures above 3,500 pounds per square inch, or more than 200 atmospheres.

That is a good deal more heat and pressure than drove the first steam engine. History dates the industrial revolution from the first successful efforts to transform heat into work—to capture fire in an engine. The object of building engines is to convert the thermal energy of fire into useful mechanical work. We know that the energy of a gas increases steadily with increasing temperature. For the purpose of building engines, only the part of the energy expressed as motion through space of the individual gas molecules need be considered. The higher the temperature, the faster the molecules move. The motion is random; the molecules fly about in all directions, colliding with one another and bouncing off the walls of the container.

The first engineers had no understanding of the molecular nature of the force they were seeking to harness. They had to proceed in ignorance of the relations between temperature, volume, pressure, and the mass of the gas present, matters that were not settled until 1800. They did understand, thanks to Galileo, Evangelista Torricelli, and, later, to Robert Boyle, that the atmosphere has mass and exerts a force (atmospheric pressure). They also knew that when steam condenses into liquid water, there is a very large volume decrease—

about a thousandfold. Boyle developed a systematic way of thinking about these matters just as the first steam-driven devices began to appear in inventors' workshops.

At first, the steam engines made use only of the partial vacuum created in a closed working vessel by condensing steam. Valves made it possible to use the difference in pressure to draw water up a pipe, as in Savery's engine. The idea of using a sliding piston was proposed by Denis Papin in about 1690 and perfected by Thomas Newcomen about 20 years later. Here in the up-and-down motion of the piston the engine secured the conversion of the random motion of the molecules in the steam into directed motion. Steam in Newcomen's engine did the relatively easy work of raising the piston and thereby lowering the working arm of the machine. Quenching and condensation of the steam then created a partial vacuum in the cylinder against which the ambient air pressure forced the piston down and raised the working arm.

James Watt's engine, at the end of the century, harnessed the positive pressure of steam to do the work. In the final "double-acting" version, steam admitted alternately at either end of the cylinder pushed the piston alternately both ways. Watt solved a host of other problems to make his engine really work: he invented the separate condenser to save heat while creating a vacuum to facilitate the return working stroke of the piston, he adapted the flyball

The steam engine designed by Thomas Savery in the late 17th century to pump water from mines had steam as its only moving part. In the first part of the cycle steam at high pressure filled the vessels. Thereupon condensation of the steam created a vacuum, permitting atmospheric pressure to push water up into the vessels. With the water-admitting valve shut, steam introduced again into the vessels pushed the water out, and the cycle would proceed. For lack of materials to contain high-pressure steam, the system never worked for mines; at smaller pressures it did serve to pump water into low reservoirs and fountains.

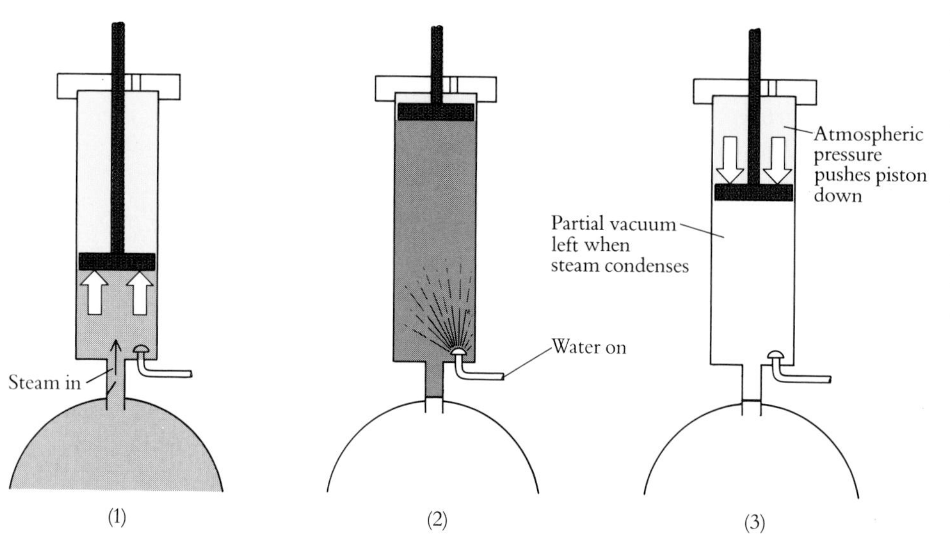

The Newcomen steam engine cycle employed steam pressure to lift the piston and raise the working arm of the pump. With the steam valve shut, cold water condensed the steam, creating a vacuum in the cylinder. Atmospheric pressure pushed the piston down in the working stroke of the cycle.

governor to control the speed of the machine, and he dreamed up several elegant ways to convert reciprocal motion into smooth rotary motion. These could be seen, along with his double-acting piston, at work into the present century in the last stationary reciprocal steam engines, and they remain at work in the steam locomotives still being built in the People's Republic of China.

Watt actually had some notion of the immense social and moral as well as practical import of his work. With other members of the Lunar Society, which

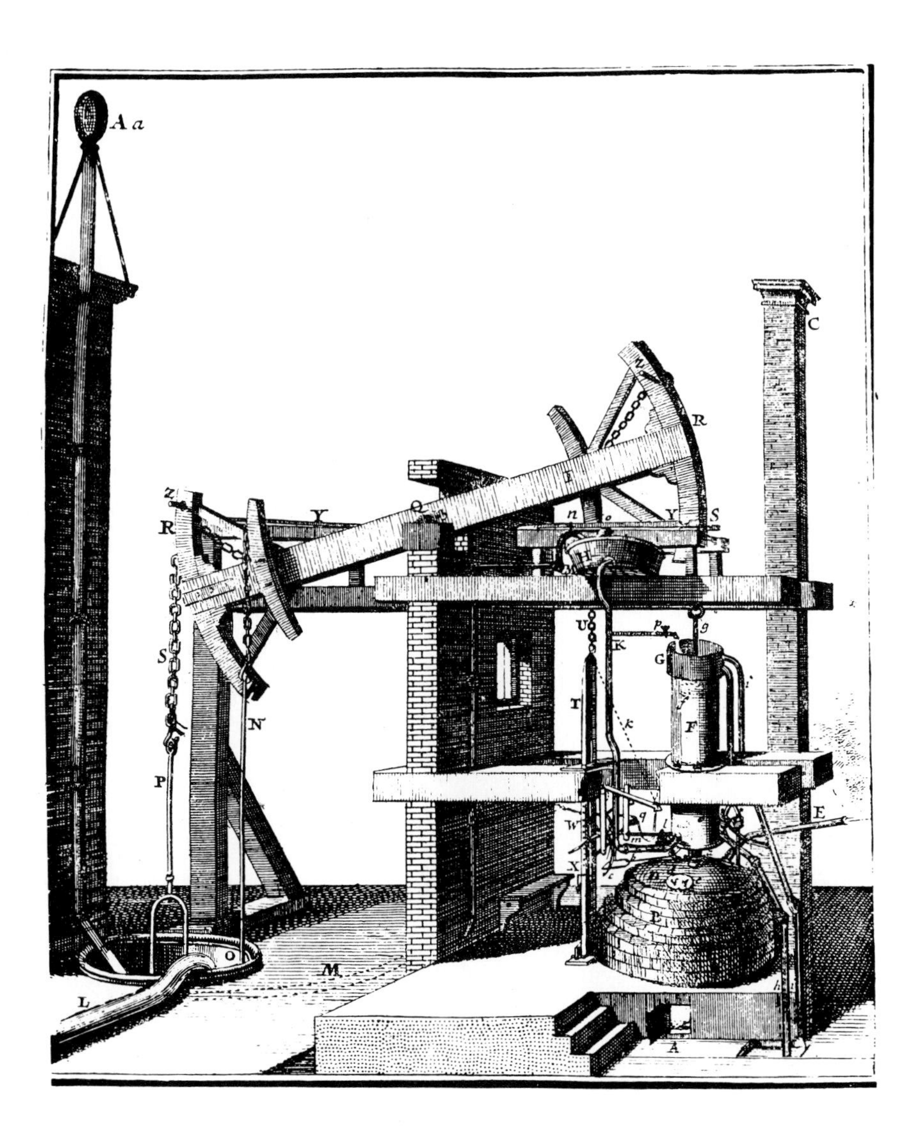

The Newcomen steam engine harnessed the weight of the atmosphere to do its work. Condensing steam in the cylinder created a partial vacuum under the piston. These engines pumped water from the mines in England for 50 years before they were replaced by Watt's double-acting steam engines.

James Watt

As his father's once-prosperous merchant business fell into ruin, James Watt at age 19 traveled from Scotland to London, where he was apprenticed to a scientific instrument maker. Returning to Glasgow in 1756 as a skilled craftsman, he began work making mathematical and experimental devices at the University of Glasgow. There he found a friend and teacher in Joseph Black, who had made substantial progress in understanding the concept of heat and the differing heat content of different materials. Black taught Watt the difference between the amount of heat present in a body and the temperature or hotness of the body. With Black, Watt studied the quantities of heat required to effect phase transitions such as the melting of ice and the vaporization of water. Watt was thus prepared to approach the steam engine from fundamental ground.

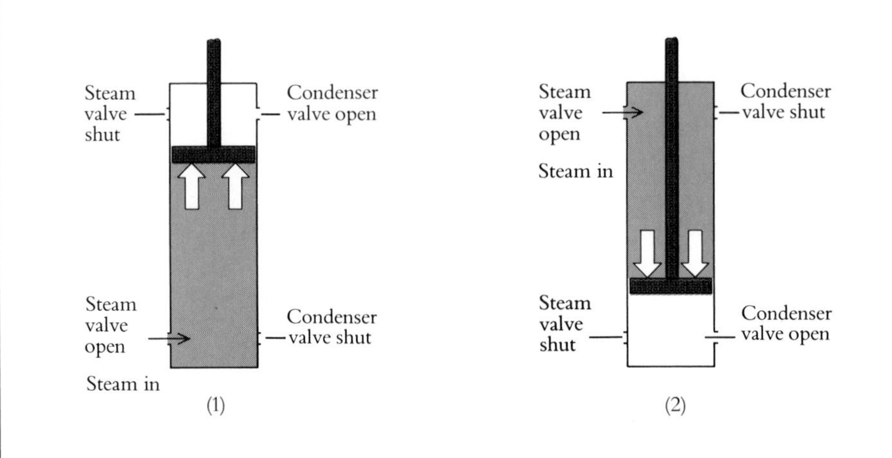

Watt's double-acting steam engine employed positive pressure of the steam to push the piston alternately up and down. This required mechanical linkage of valves at either end of the cylinder.

Conversations between Watt and Black often turned to the topic of steam-engine design, and Watt once happened to be given the task of repairing a model of Newcomen's engine used in university classes. Studying the model as he worked on it, and

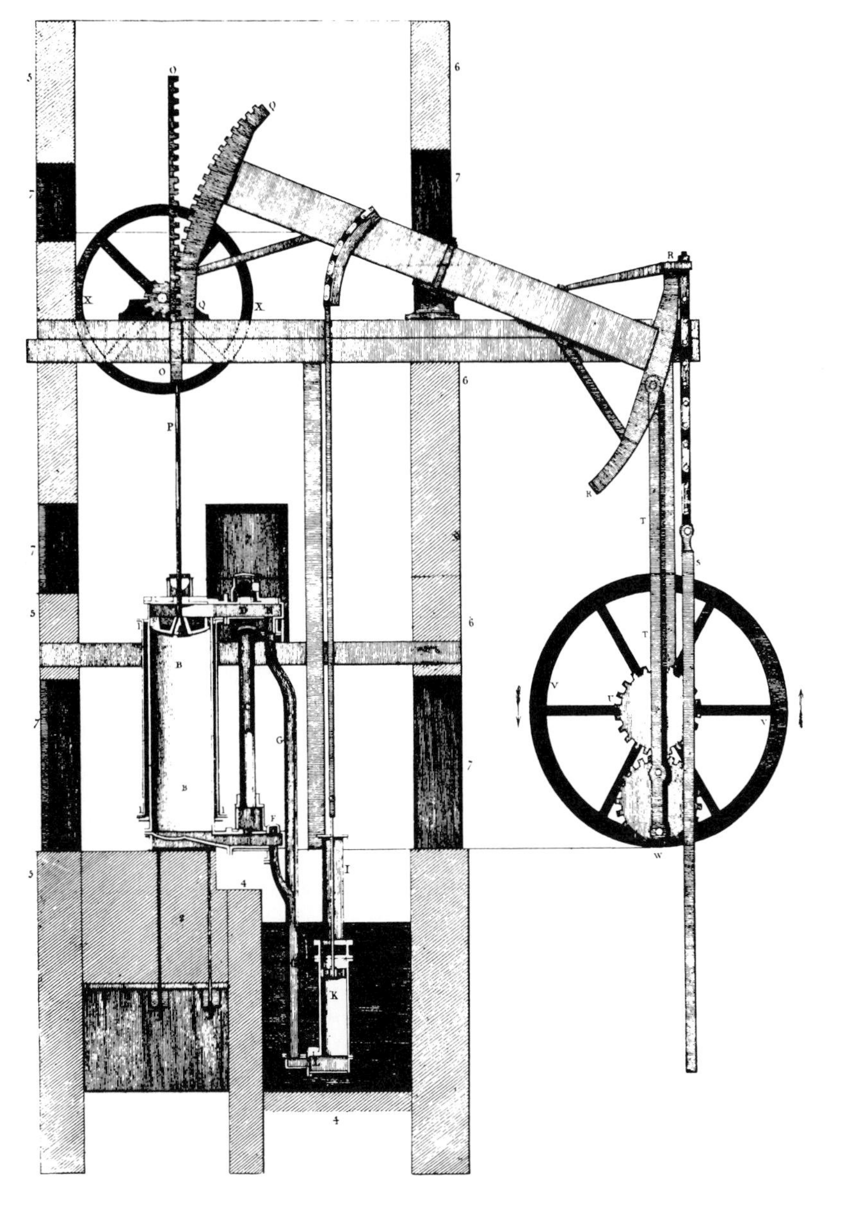

Watt's own design for the installation of one of his engines suggests the complexity of the mechanical linkages that made his double-acting steam cycle possible. Valves at the top and bottom of the cylinder (B) had to be opened at just the right times to admit steam alternately to the two sides of the piston. The separate condenser is shown at K. The planetary gear attached to the right end of the rocking arm translated its oscillatory motion into the rotary motion of the big flywheel.

applying his knowledge of heat phenomena, Watt realized that the Newcomen cycle made inefficient use of the total available energy in the steam, merely using it to push the piston up and raise the working arm and using the atmosphere to push the arm down on its work stroke—and then threw much of the heat away by cooling the cylinder immediately. This insight formed the basis for all that was to follow. Watt first determined to keep the cylinder and piston hot by physically moving the condensation process some distance, putting it in a separate device called a condenser. This brought about a great saving in heat. He subsequently figured out how to extract additional work from the steam in his engine by allowing the volume to expand while reducing the pressure in the cylinder, and by inventing the double-acting steam engine: it admitted steam alternately to each side of the piston. Given the difficulty at that time of acquiring accurately machined parts and making tight seals, these were dramatic advances. Indeed, the success of the steam engine and its ramifications in the new manufacturing industries that drew on its power pushed the production, fabrication, and machining of metals to a new high rate of progress.

Watt obtained patents to protect his inventions, and he traded a share of his profits from the patents to obtain financing from successful businessmen. Eventually he formed a company in partnership with Matthew Boulton to design, manufacture, and sell steam engines. Watt concentrated on engine design while Boulton handled the business. The engines produced in Soho by Boulton & Watt soon replaced the Newcomen engines in mines throughout Britain. The new engines used only about one-fourth as much fuel as the Newcomen engines, and Boulton devised the very successful approach of selling an engine not for a fixed price but for one-third of the fuel savings that the engine actually provided.

Watt carried on other research in addition to that aimed solely at improvements in the steam engine. He spent some time in the 1780's studying the composition of water, demonstrating that water is a combination of hydrogen and oxygen (or, as they were known at the time, phlogiston and dephlogisticated air). He even showed that the volume ratio of these two components in water is 2:1. His findings were communicated to a number of chemical researchers across Europe, and Antoine Lavoisier incorporated this knowledge into his synthesis of the new chemistry. In 1800, Watt turned his share of the company over to his sons and devoted the rest of his life to research and invention.

Watt was so far ahead of everyone else in his understanding of the steam engine that no one was able to extend his work for many years after his death. Indeed, there was a noticeable decline in the efficiency of steam engines for many decades after.

counted among its members Erasmus Darwin, Joseph Priestly, Josiah Wedgewood, and Watt's financial backer Matthew Boulton, he met each month over the years from 1764 until 1780 to discuss these implications as well as the substance of natural philosophy. Only partly in jest, the Scottish science journalist Ritchie Calder argued that the industrial revolution was a conspiracy. Certainly it was a revolution. It has been shown that around the time when Abraham Lincoln issued the Emancipation Proclamation in 1863, the rising curve of mechanical horsepower from steam engines had crossed the output of work from human muscle in America.

In the generation of electrical power, the reciprocal steam engine yielded in the last years of the 19th century to the steam turbine perfected by Algernon Charles Parsons. It is these machines that really challenge boiler engineers to raise the steam in their furnaces. With steam at an inlet temperature of nearly 600°C and pressures of up to 300 atmospheres, a steam turbine today will generate as much as 75,000 horsepower, its immense mass spinning at 10,000 rotations per minute. The steam impinges on the first buckets of the turbine almost as a plastic solid and yields its energy to the rotary motion of the turbine as it expands through 20 or more stages into the exhaust steam chest.

The steam turbine is the principal prime mover in the conversion of heat into mechanical energy in the generation of the world's electrical power.

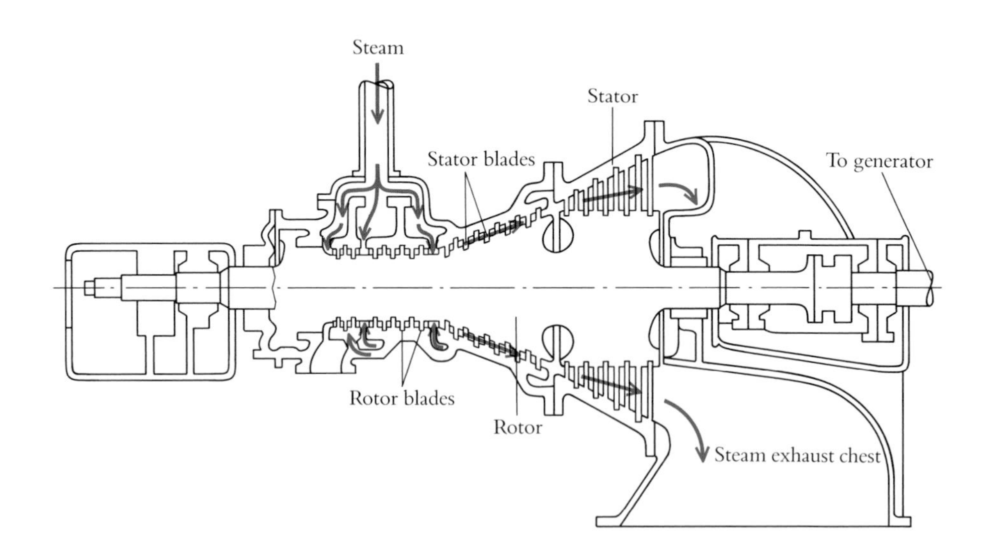

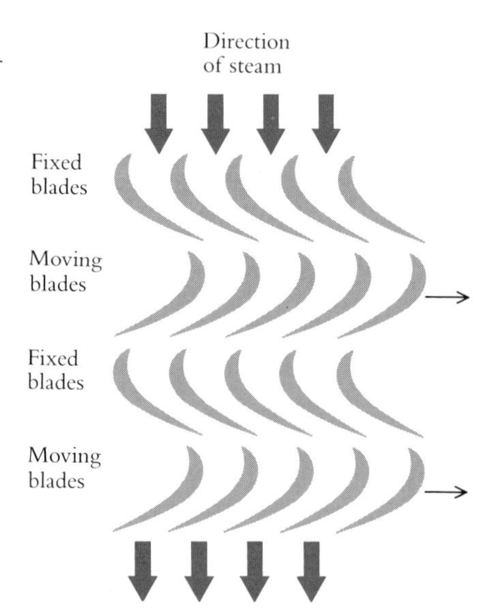

A steam turbine converts the linear flow of steam into rotary motion much as a windmill captures the energy of wind. The rotor, the principal moving part, bears many rows of buckets, or blades, standing out at right angles to its axis. Each row is a "steam wheel"; the succession of wheels extracts the energy of the steam, step by step. Ahead of each wheel of blades in the rotor is a fixed, stationary wheel of buckets mounted in the stator, which houses the rotor, mounts its bearings, and contains the steam. The space between each pair of blades in the "reaction turbine" schematically diagrammed here forms a kind of nozzle through which the expanding steam accelerates (see the cross-sectional diagram of blades at right; black arrows indicate the direction of motion of the blades). Through the nozzles of the stationary blades, the steam jets into the nozzles of the moving blades. The jetting of the steam from the nozzles of the moving blades imparts by equal and opposite reaction the force of its motion to the rotor. Entering the turbine at left, the steam expands through the succession of the turbine wheels from left to right, losing temperature and pressure to the energy of motion it imparts to the rotor until, greatly expanded in volume, it flows into the exhaust chest and out of the turbine. The shaft of the turbine turns the shaft of an electrical generator, which, in continuous performance of Faraday's induction experiment, converts the energy of the rotary motion into electrical energy.

It occurred to inventors early in the history of the steam engine that, in place of steam, they might make fire itself their working fluid. In fact, more than 30 years before Newcomen, Jean de Hautefeuille set off gunpowder under a piston to raise it up and then cooled the cylinder to help the weight of the atmosphere push the piston down. The first attempt to burn a fuel in a cylinder came in 1820, when William Cecil burned a hydrogen-air mixture to raise the piston, depending, again, on the atmosphere to bring it down. In 1862 Alphonse Beau de Rochas set down the principles of the four-stroke-cycle internal combustion engine. The names of Nicholas August Otto and Eugen Langen are attached to that cycle in testimony to their success in 1878 in making the four-stroke-cycle system work. What made this development possible was, of course, not only the inspiration of engineers but progress in metallurgy and the blooming of the technology of electricity from Faraday's work in the first third of the century.

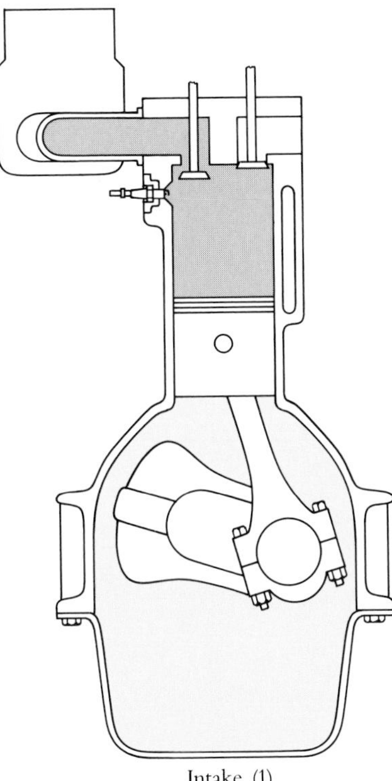

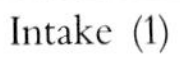

Intake (1)

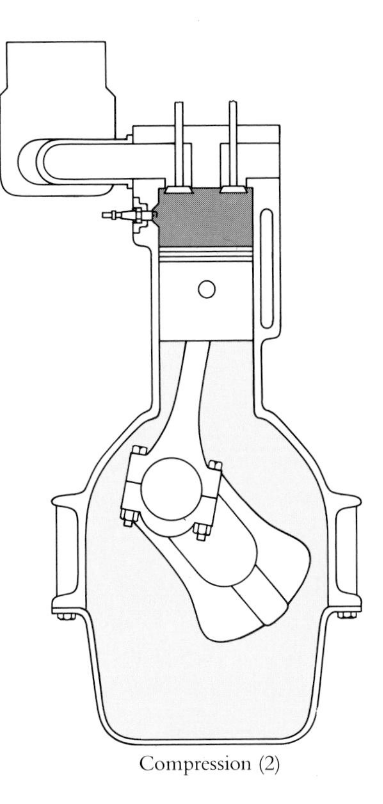

Compression (2)

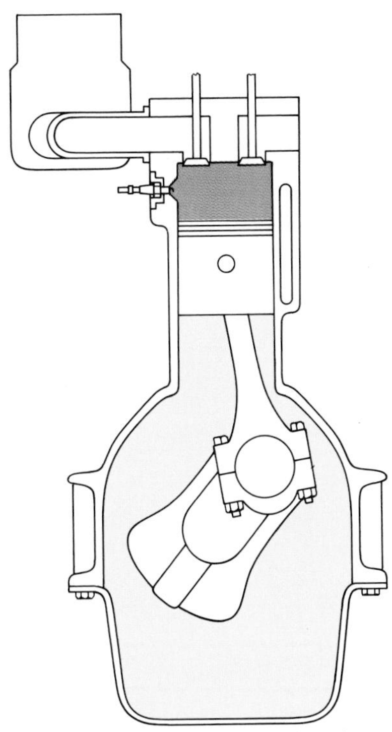

Expansion (power stroke) (3)

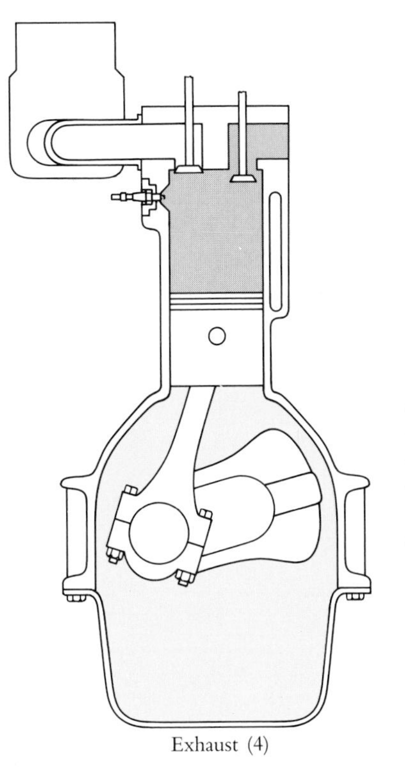

Exhaust (4)

The four strokes of the conventional internal combustion engine are shown. In the first revolution (here clockwise) of the crankshaft on the intake downstroke, the fuel valve opens to admit fuel to the combustion chamber; the return upstroke, with the fuel valve shut, then compresses the fuel-air mixture. Ignition of the fuel, just before the completion of the compression stroke, starts the combustion of the gases that pushes the piston down again in the power stroke. The return upstroke of the piston in this second revolution of the crankshaft drives the combustion gases out of the cylinder through the open exhaust valve.

It is the burning—burning, not detonation—of the fuel in the combustion chamber that powers the engine. This occurs over a 50-degree change in the crankshaft angle, a short but far from instantaneous event. This is almost one-sixth of the 360-degree cycle of the working—compression and power stroke—half of the total 720-degree cycle. Burning begins with a fast first stage in which the combustion reactions are initiated by an electric spark and occurs best with the fuel-air mixture a little fuel-rich at the peak of compression. The second stage is travel of the flame front throughout the enlarging combustion chamber. That travel should proceed smoothly enough to exert a steady push on the retreating piston.

Engine knock, which interrupts this smooth expansion of the gases and wastes some of the power of the enlarging flame front in noise (the familiar "ping"), vibration, and heat, is not a premature detonation as once thought. It is rather autoignition of a portion of the fuel before the flame front proper arrives. The mixture is heated to ignition by a pressure wave ahead of the flame and by radiation from it.

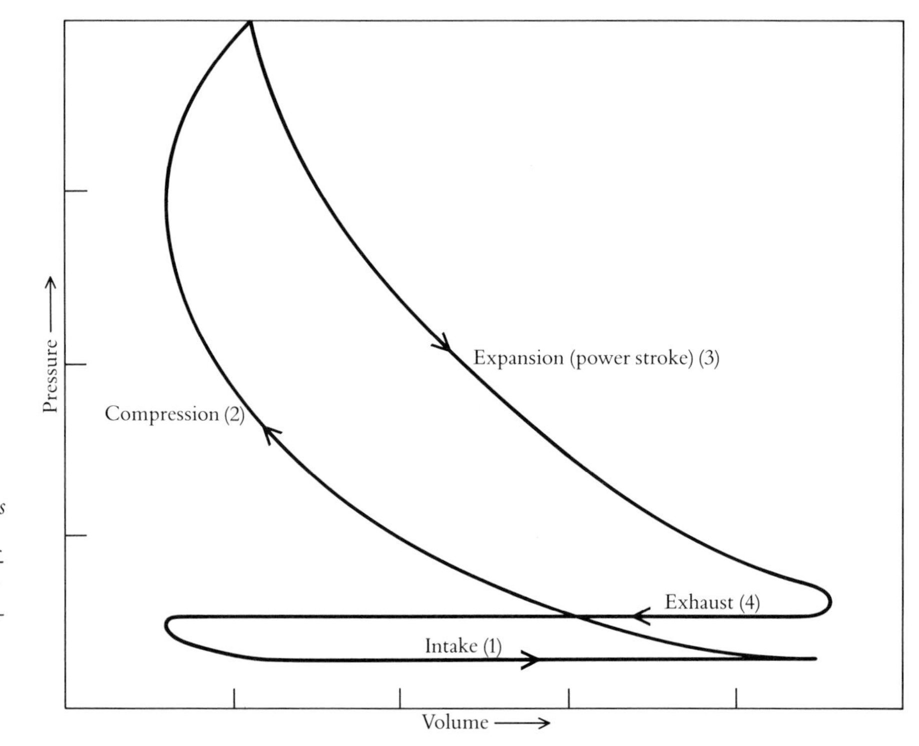

The curve in this chart shows the relation of pressure to volume in the cylinder of the internal combustion engine through its four cycles. With the fuel valve open in the intake downstroke (1), pressure is low and constant as volume increases. In the return compression upstroke (2), volume decreases and pressure increases to the maximum. In the power downstroke (3), pressure decreases with increasing volume, but remains higher than during the compression stroke. The difference in pressure between these two curves is the contribution of the heat of the combustion of the fuel and constitutes the driving force of the system. On the return upstroke (4), with the exhaust valve open, pressure remains constant as volume decreases.

The susceptibility of the combustion process to knock increases with compression. Increase in compression, however, is the road to efficiency. A rough predictor is:

$$\text{Efficiency} = 1 - \frac{1}{r^c}$$

where r is the compression ratio and c is a constant, with values of 1.2 to 1.3.

Higher-compression engines are more efficient or, alternatively, the same engine delivers more power at higher compression. The fact that the exponent c is greater than unity pays a premium extra increase in efficiency for any increase in compression. Typical gasoline-powered spark-ignition engines operate at a compression ratio of 1:7 to 1:11, the ratio being that between the volume of the cylinder at the top and at the bottom of the stroke. The temper-

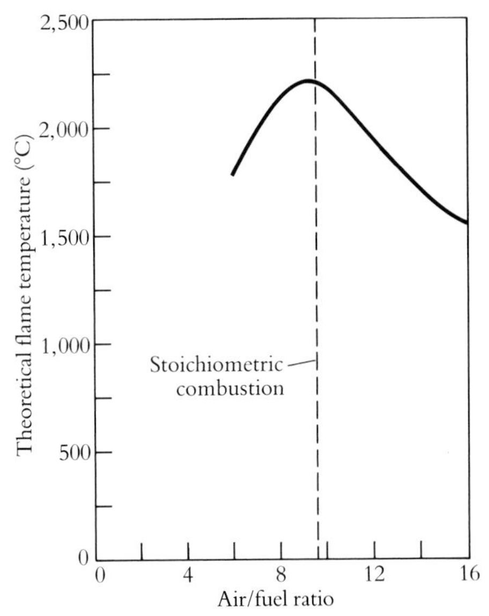

An increase in the air-to-fuel ratio secures an increase in the temperature of combustion up to the point where there is sufficient oxygen to consume the fuel; thereafter the flame temperature is cooled by the excess of air (including its excess of oxygen). On this ideal curve, the stoichiometric combustion point, with a slight excess of oxygen, proves most efficient in practice.

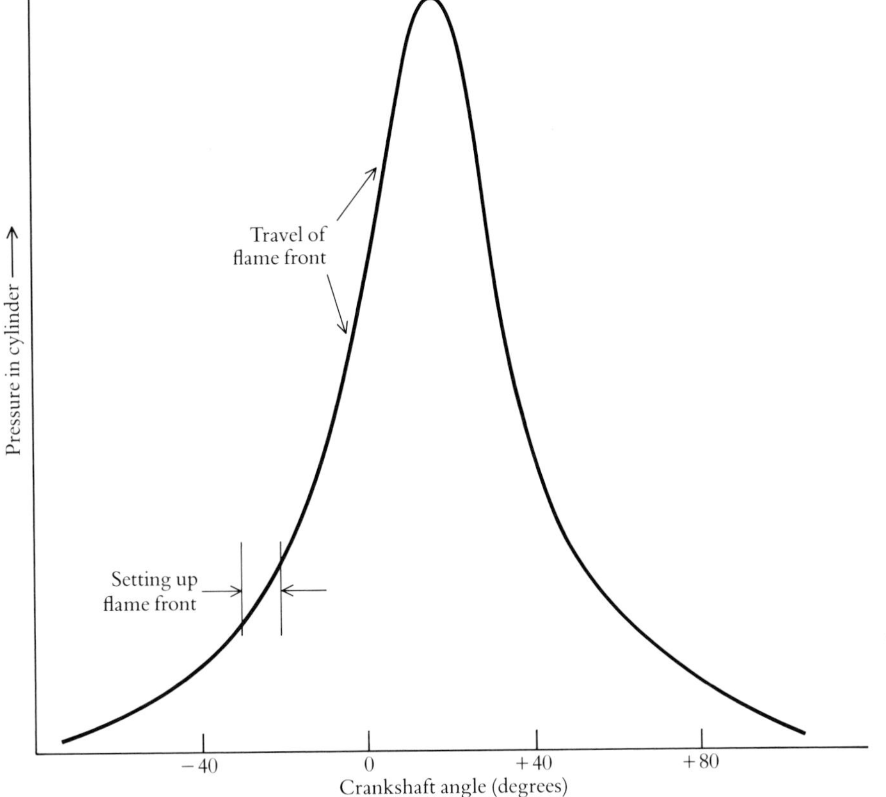

Change of pressure in a combustion chamber is plotted against change in crankshaft angle. Pressure starts rising with the crankshaft approaching an angle of −40 degrees on the compression upstroke and then rises steeply with ignition just before completion of the upstroke (0). Pressure continues rising on the downstroke until the crankshaft turns to an angle of +20 degrees. The pressure then falls as the piston retreats down the cylinder.

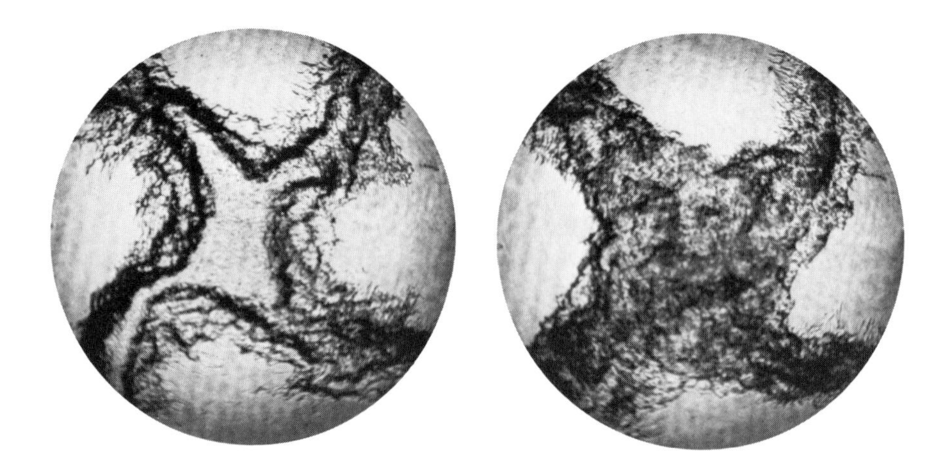

A photograph taken through a quartz head on the cylinder of an experimental internal combustion engine contrasts smooth combustion at 24.8 bar pressure (left) with premature ignition of the fuel ahead of the flame front, producing "knock" at 26.0 bar pressure (right).

ature of the gases rises rapidly with the compression ratio until at 1:15 (a pressure of over 600 pounds per square inch) the temperature reaches its autoignition point (550°C). Engines running at these compressions—that is, diesel engines—need no spark plugs. For the spark-ignition engine, the top compression must remain below the autoignition point, the practical limit being near 1:11.

In the immediate aftermath of World War II, when the U.S. automobile industry committed itself to big cars and high compression, the knock problem came to the fore. With fuel a minor consideration in those days of petroleum plenty, the industry elected to cure knock by chemistry. It adopted high-octane gasoline. Among the hydrocarbons in gasoline are fairly heavy ring structures in the benzene and naphthalene families and light molecules such as the isomers of butane and pentane. The volatility needed for good performance in cold weather is provided by light fractions. The knocking behavior is scaled on the performance of isooctane:

$$\begin{array}{ccccccccc} & & CH_3 & & & & CH_3 & & \\ & & | & & & & | & & \\ CH_3 & — & C & — & CH_2 & — & CH & — & CH_3 \\ & & | & & & & & & \\ & & CH_3 & & & & & & \end{array}$$

arbitrarily assigned a performance number of 100, and normal heptane:

$$CH_3—CH_2—CH_2—CH_2—CH_2—CH_2—CH_3$$

which is arbitrarily assigned a rating of zero.

The U.S. automobile and fuel designers added tetraethyl lead to the fuel as a knock suppressant. This additive somewhat alters the chain reactions in the combustion front to make autoignition more difficult without affecting the burning rate, when ignition is with a spark plug.

Today concern for the effects of lead in the environment and on public health, combined with the high price of petroleum, puts the chemical solution of the knock problem in doubt. An alternative mechanical cure, available all this time, lies in the design of the combustion chamber. Thus the so-called stratified-charge combustion chamber in a Japanese automobile engine establishes a fuel-rich zone around the spark plug to accelerate the first stage and a leaner mixture elsewhere. Enhancing the mixture of fuel and oxidant by turbulence speeds up the second stage. Other solutions are to use a second spark plug and additional inlet valves.

The amount of air required to burn gasoline is surprising. It takes, in round numbers, four units of oxygen to burn one unit of gasoline. Since oxygen makes up 0.21 percent of air, nearly 19 units of air are needed for each unit of gasoline. That calculation reckons with gasoline as a vapor at sea-level pressure and room temperature. With gasoline as a liquid, one gallon requires 2,240 gallons of air, or about 360 cubic feet! That suggests the magnitude of the volumetric expansion of the fuel into hot gases by combustion that powers the automobile.

The diesel engine, with its high-precision fuel injectors and its high operating pressures, is a more expensive machine. It ultimately replaced the steam engine in the ships of the world and on the railroads in the United States, and it powers the long-haul trucking system in this country. In the private automobile market its higher cost and somewhat sluggish performance outweigh its economy and efficiency.

In thermal efficiency—the net work derived divided by the net heat put in—both the gasoline and the diesel engine outclass steam:

Steam	.30
Gasoline (spark-ignition)	.60
Diesel (compression-ignition)	.55

These figures are substantially reduced by friction losses and the need for power accessories. At the wheel the efficiency of spark-ignition engines is reduced to about .30 and the diesel to a somewhat better .35.

The reciprocal engine never made a wholly satisfactory power plant for aircraft, whether its cylinders were arrayed in line or radially. Air transportation could not have secured its commanding place in the travel of people and its superlative safety record without the gas-turbine jet engine. The advent of that engine, in which the combustion gases again replaced steam as the work-

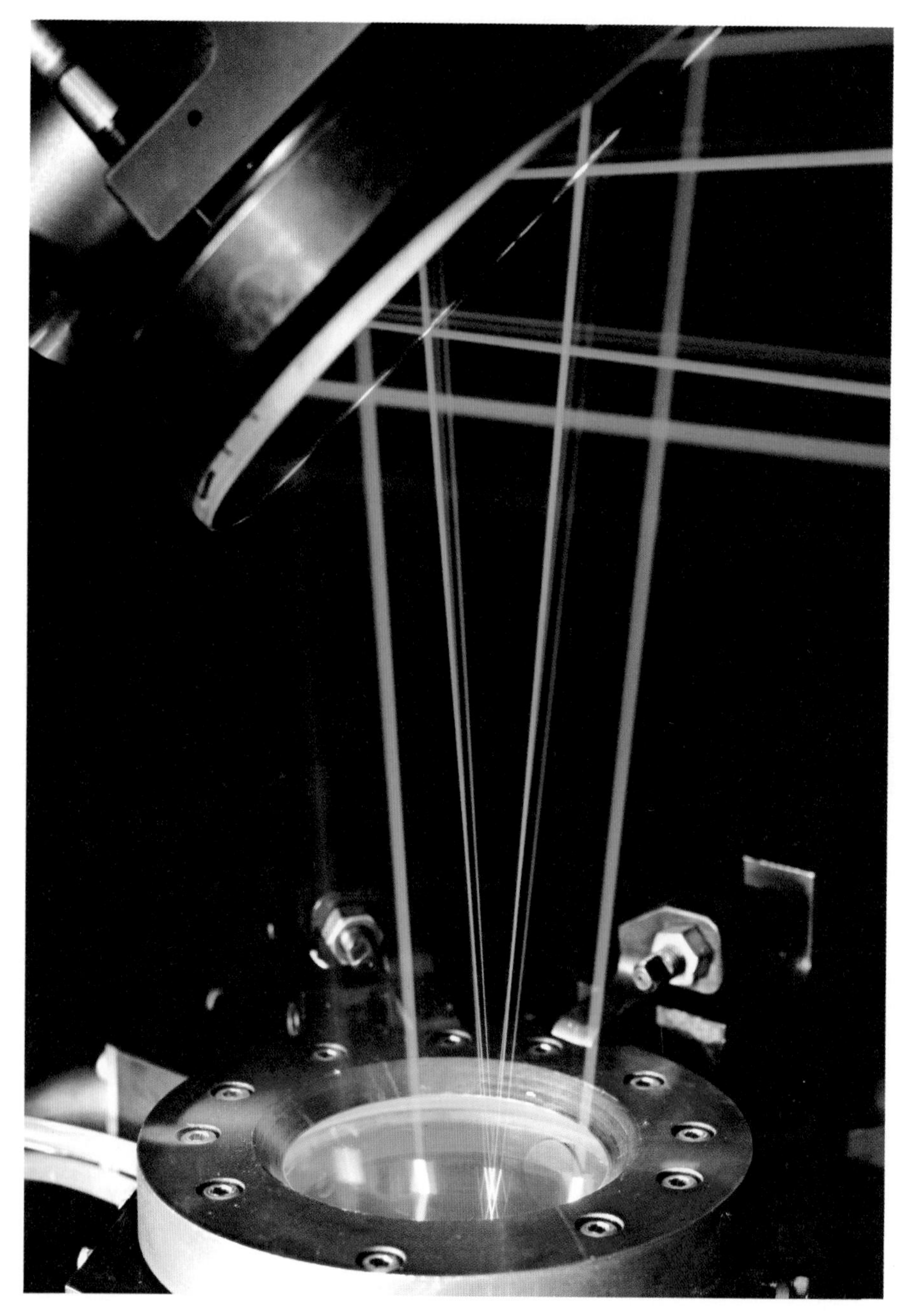

Laser-beam spectroscopy (see the figure on page 32) is employed here to identify intermediate combustion compounds formed in the flame front in a combustion chamber fitted with a quartz head, in pursuit of a chemical cure for knock at high pressures.

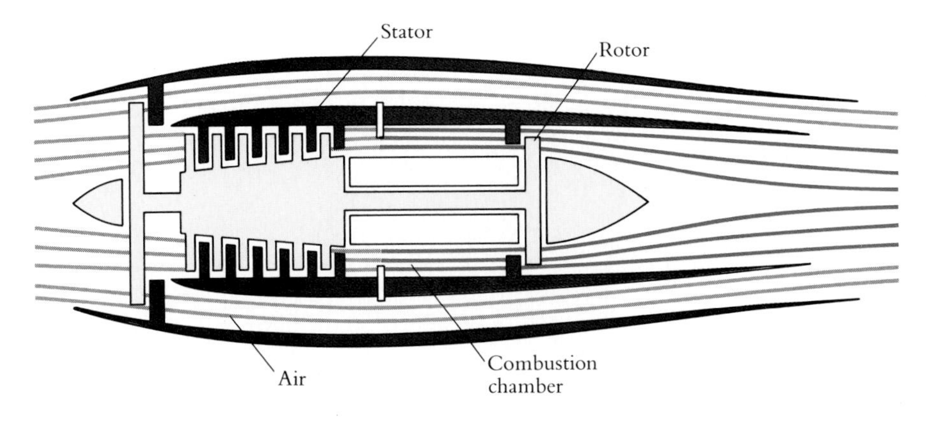

A gas turbine for aircraft propulsion achieves high horsepower or kilowatt rating per pound of engine by an enormous energy throughput. The single-stage turbine at right drives the multistage compressor at left to deliver a large mass of air at high pressure and velocity into the combustion chamber, from which the combustion gases, at still higher pressure and velocity, drive the turbine. The turbine delivers energy to spare to drive the single-stage compressor fan ahead of the turbine, which in turn drives another large mass of air around the engine and aft, into or with the turbine exhaust stream, augmenting the thrust of the expanding exhaust from the turbine.

ing fluid (agent for transmitting energy), had to await advances in metallurgy and feedback control.

With a hot-gas turbine fixed on the same shaft, a compressor delivers air to the combustion chamber. Exhaust gas from the combustion chamber drives the turbine, which at once operates the compressor and delivers the thrust of the hugely expanded and accelerated combustion products to the aircraft. The temperature rise in the combustion chamber is impressive, on the order of 500 to 600°C. The heat-release rate of these engines is considerably larger than that of big public-utility boilers. It is the high ratio of power derived per unit engine weight that makes the gas turbine the power plant of choice for aircraft.

The gas turbine drives the reactions in the combustion chamber to their limits. Gas velocities, of both the incoming air and the necessarily matching velocities at which fuel must be delivered, tend to exceed the velocity of their combustion in the flame. The flame threatens to fly downwind. Swirling of the airstream is one technique to manage this problem; another is to alter the streamlines dramatically, even to reverse the direction of flow, and bring the gases back into the combustion chamber proper.

These problems have been managed well enough so far to bring in the turbofan, almost a return to the propeller. The turbine on the primary circuit takes off just enough energy to drive the compressor serving the combustion chamber. The remaining energy drives a second, larger-diameter turbine wheel that is coupled to a large-volume, moderate-pressure compressor called the fan. The fan gathers up and drives a large volume of compressed air to a main expansion nozzle at the rear of the engine. The hot exhaust from the engine may also exit through the same expansion nozzle or exit in parallel. The additional mass of air thus entrained substantially increases the total thrust delivered by the engine.

Saturn C-5 engines lift the Apollo 11 spaceship from Earth on the first lunar landing mission, July 16, 1969. The five regenerative liquid-fuel rocket engines expended in 150 seconds enough energy to light a city of 100,000 for a day.

The gas-turbine aircraft engine must be regarded as a superlative machine for the speedy and efficient transformation of heat into mechanical work. In this respect, it commands the admiration of the steam-boiler engineer. It is exceeded in the speed at which it accomplishes this transformation only by the pure reaction engine of the rocket. The Saturn C-5 engine, for example, burns liquid hydrogen and oxygen at the extraordinary combustion-chamber temperature of about 3,000°C. For the brief duration of the "burn" the chamber walls are kept in the solid state by the cooling inflow of the fuel and oxidizer gases pumped, heat-exchanger style, through its skin. For each of those 150 seconds after ignition each of the five engines delivers 1,600,000 pounds of thrust.

Kilowatts, for power or rate of doing work, and kilowatt-hours, for total energy output or work done, offer common denominators for comparing the performance of the three remarkable heat engines we have been considering here. The 75,000-kilowatt steam turbine generates about two million kilowatt-hours a day. That is enough to light a city of 100,000 population and get its day's work done. On the nonstop flight from New York City to Tokyo, the four gas turbines on a 747, with a combined power of 150,000 kilowatts, expend the same amount of energy in half a day. To put a payload into orbit the five-engine power plant of the Saturn C-5 rocket, with a capacity of 5.6 million kilowatts, delivers more than two million kilowatts in 2.5 minutes.

This is a measure of the fire force that escapes sometimes as violence to wreak havoc on life and property.

An experimental fire conducted by Professor Howard Emmons at Harvard University and his colleagues at the Factory Mutual Research Corporation demonstrates the standard course of an unwanted fire in a room. Through the doorway at the left, the first flames have come up at two minutes on the clock. The numbers on the walls permit measurement and timing of the growth of the layer of hot combustion gases and soot under the ceiling. Instruments for measuring temperature and sampling the atmosphere are arrayed across the door at regular intervals between the floor and ceiling. At four minutes, a thick layer of hot gases has formed and is just coming out of the room at the top of the doorway. At six minutes the layer is so deep it pours out under the top of the door; the sharp boundary at its bottom also defines the top of the layer of inflowing cool air. At five seconds past seven minutes, the whole room goes into flames: flashover (see the figure on page 87).

5 *Unwanted Fire*

As the start-up of the flame of a candle or the lighting of a fire in the fireplace proceeds from its beginning in accordance with a scenario, so does the course of an unwanted fire. The fires that concern us here—those in U.S. households, which exact most of the loss of property and life—start small. They start smaller today than in the years when flames from candles, lamps, stoves, and hearths presented the greatest fire dangers in the home. Most domestic fires these days are started by cigarettes and malfunctions in electrical devices and wiring.

The fire that burns down a house first burns up one room. In that room it starts small. It goes through a phase of modest growth. Then, in a few moments of very rapid acceleration, it engulfs the entire room and every combustible thing in it.

People are adjusted to linear phenomena. Caught in the first phases of a fire, they expect it to grow steadily and predictably. They are often overtaken by the rapid buildup to flashover: the sudden transition to full involvement of all combustibles in the room. The moment of flashover is striking and startling even for one who has seen it many times, as I have in experimental fires.

It is worth examining events in a room fire to see how flashover comes to happen.

All too commonly a fire begins when a cigarette drops onto upholstery or into bedclothes as a person falls asleep with a burning cigarette in hand. A cigarette, of course, is designed to sustain steady smoldering combustion without bursting into flame. It starts similar smoldering combustion where it drops. Sometimes it continues to smolder for hours, filling the house with toxic vapors. At other times the smoldering changes into flaming combustion. With an open flame to heat up adjacent objects to their ignition temperatures, the second phase—the phase of exponential growth—begins.

The heat from a fire is proportional to the area burning. The more area aflame, the more heat is liberated; the growth rate is proportional to the size of the fire. In a dimensionless mathematical equation, this statement reads:

$$\text{rate of growth} = k \times \text{present size}$$

A smoldering fire in an upholstered chair is started by a cigarette in a crevice between the seat and the right armrest, at left. At 15 minutes, center, the fire has crept up the armrest, charring the fabric. At 40 minutes, right, the armrest is charred to the top and smoke obscures the scene.

where k is a constant, or

$$\frac{\text{change in size}}{\text{change in time}} = \frac{\Delta\text{size}}{\Delta t} = k \times \text{size}$$

or

$$\frac{\Delta\text{size}}{\text{size}} = k \times \Delta t$$

This relation of small changes, or differentials, may be expressed in integral form:

$$\text{size} = Ae^{kt}$$

where A is another constant.

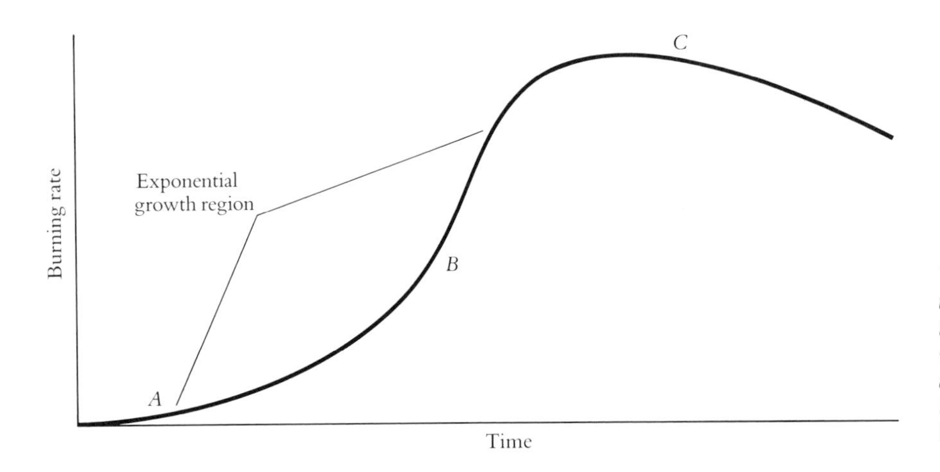

The burning rate of a fire plotted as a function of time increases slowly at first and then exponentially through the flashover, which sets every combustible in the room aflame. The curve then levels at a steady burning rate and finally falls off as the fire burns out.

The equation, plotted as part of the bottom figure on page 86, says that the production of heat and of other combustion products increases exponentially with time. The development of the fire during this phase is proceeding up the steep slope of the curve shown in the figure toward the moment when it becomes a room full of flames. Decisions made during this time are life-and-death choices.

The rest of the room, even distant from the flames, is being heated, mostly by black-body radiation from the luminous flames, made luminous, as is the candle flame, by incandescent soot particles. Many of them go uncombusted into the smoke. A layer of hot gas and soot from the incomplete combustion in the fire—combustion in an unwanted fire being not as complete as in a furnace—is building up under the ceiling, and this is radiating its own heat. The temperature of the layer of gas and soot approaches 500°C. Needless to say, this atmosphere is lethal: body temperature is 37°C. The floor under the direct radiation from above is heated to 300°C or so. The air near the floor, however, is largely incoming air from outside and is not heated rapidly by radiation from above. Hence in this situation it is advisable to keep one's head low to breathe the cooler air while escaping.

Wisps of steam or smoke rising from objects and surfaces distant from the flames give the telltale sign that the rest of the room is heating up to the ignition point. The hot gas and smoke layer under the ceiling has deepened enough for there to be substantial flow out through the upper half of the doorway or the windows. Any such sign means the fire is about to grow explosively.

Upon flashover every surface and object erupts in flames. The curve plotted in the figure on page 86 now levels out, for the fire is burning over a constant area; the rate is limited by the processes involved in burning in depth.

An experimental room fire is shown here through three phases. The small fire in the top illustration grows in size and intensity; its radiation begins to heat distant objects in the room. Heating of the room now proceeds exponentially—with increasing acceleration—as the hot, sooty smoke layer builds up under the ceiling and contributes its radiation to the heating process. In the flashover at bottom, with all surfaces in the room heated to the ignition point, everything in the room goes simultaneously into flaming combustion.

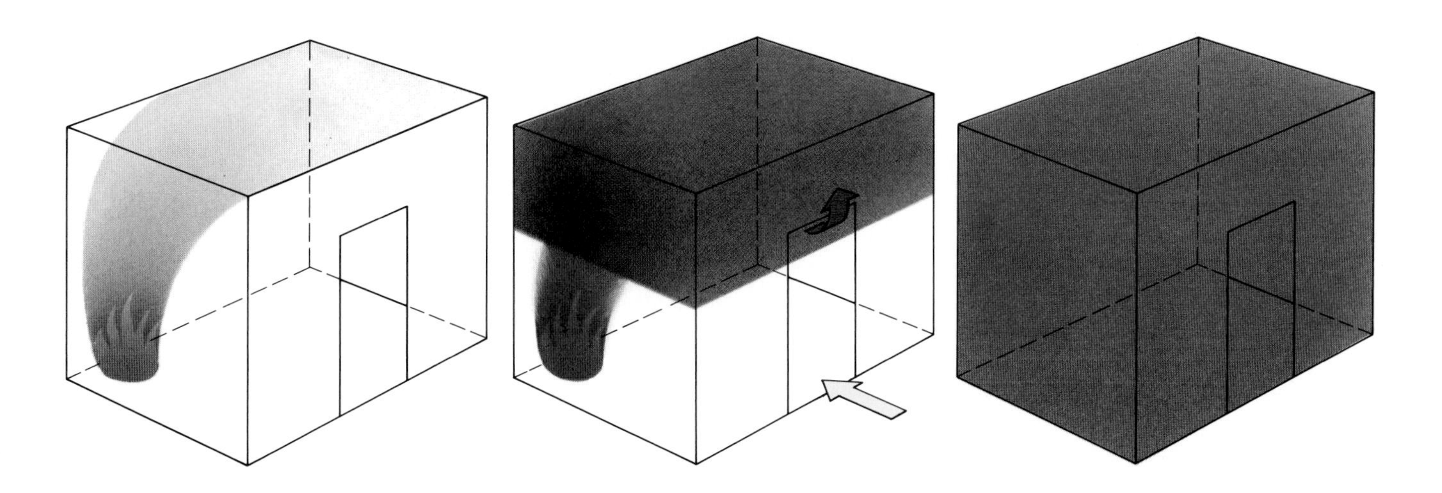

Density and Heat of Combustion of Various Materials

	Density (g/cm^3)	Heat of combustion (kcal/g)	(kcal/cm^3)
Methyl alcohol	0.8	4.8	3.8
Cellulose	0.4 (fir)	4.0	1.6
	0.7 (oak)	4.0	2.8
Polyethylene foam	0.035	9.8	0.34
Polystyrene foam	0.016	9.1	0.15
Polyisocyanurate foam	0.036	6.3	0.23
Polyurethane foam	0.035	6.2	0.22
Gasoline	0.95	11.5	10.9

Eventually the combustibles in the room burn out one by one, and the fire—if it is an experimental one—dies away.

When it comes to doing something to prevent or halt catastrophes of this kind, measures must start from the recognition that our surroundings are combustible and will continue to be so. Even though people know concrete and steel do not support combustion under ordinary circumstances, they will not live in such an ambiance, wearing asbestos clothes and sitting on ceramic chairs. The fact that the environment of the house is combustible does not persuade people to dispense with the amenity of the clothes they wear, the bedding in which they spend a fourth of their lives, the drapes at the window, the chair and desk in the den, the upholstered furniture in the living room. The gypsum materials in the wall and the ceiling will not burn, but the wallpaper will. The frame of the house is wood, the roof is wood, perhaps with asphalt shingling, and the exterior is probably wood, at least in part. The structure could, of course, be made safe from fire, as the stone, brick, and concrete of many of the multiple-occupancy dwellings in the cities are. Even if it were, the contents of the structure would remain as combustible as before.

How these combustibles behave when they are exposed to or in a fire must accordingly be of interest. Research in fire prevention pays a good deal of attention to the questions that arise here. First and foremost of the physical properties of a material regarded from this vantage is its total heat of combustion. That is the total thermal energy per unit mass that can be released by its complete combustion. The second significant property is the rate of combustion or rate of heat release of a material; this says something about the intensity of its output of thermal energy per unit mass per unit time. These we may think of as the "exothermic" properties, measuring what the material has to contribute to a fire once it is set burning. The heats of combustion of materials vary widely, differences that are intensified by differences in density. Thus gasoline has a high heat of combustion and, with a fairly high density, a high heat content per unit volume. Foamed plastics, at the other extreme, have roughly the same heat of combustion; that is, the same total thermal energy per unit mass. With the density reduced by foaming, however, they have less energy per unit volume to contribute to the fire.

What may be called correspondingly the "endothermic" properties come next in interest: what it takes to sustain the burning. The properties to be considered are heat capacity, sensible heat, and latent heat. The heat capacity of a material states, as for the candle, the heat required to raise the temperature of a unit of the material by one degree. The raising of the temperature encounters, at certain points on the thermometer, the need to pump in extra heat. The latent heat is the heat required to secure the phase change from solid to liquid; then, at a higher temperature, from liquid to vapor.

Most of the combustibles that burn in an unwanted fire are solids that do not melt. Their substance, cellulose for example, is decomposed by heat directly to gas. In this case there is only the latent heat of vaporization and no latent heat of melting. The process releases fragments of the constituent molecules, as in a log on the fire, to burn and perhaps to sustain the burning of the material. In this connection still another property must be considered; this is the thermal conductivity of the material, which is the measure of the rate at which heat will penetrate into the interior from its surface to keep the process of decomposition going.

When these two sets of properties are summed for all the materials in a room, the surplus of heat of combustion over the inputs of energy required to bring the material to combustion gives a measure of the excess thermal energy available to feed back to the fire. The larger the difference, the more such energy is available to sustain the fire and spread it elsewhere.

Such an assessment of physical properties goes more particularly into evaluating the fire behavior of species of materials. The contrast between materials at two extremes will serve. A piece of foamed plastic burns like a marshmallow. It has a high ratio of heat of combustion to the heat required to secure its decomposition to vapors. It also has low thermal conductivity, low density, and low heat capacity. Fire tests of spaces lined with combustible insulation often result in the full envelopment in flames of the entire space in a brief, high-intensity fire. Yet when the fire goes out, there will be intact insulation just under a thin layer of char. At the other extreme, sound hardwood furniture has a high heat capacity, high density and high conductivity, and a heat of combustion substantially greater than the heat required to decompose it. Such a material heats up slowly and uniformly and burns in depth. It is hard to get

Ratio of Heat of Combustion and Heat of Gasification for Various Materials (Heat Available to Support Further Combustion and Heating Up of Surroundings)

Material	Heat of combustion, ΔH_c (kcal/g)	Heat of gasification, ΔH_g (kcal/g)	Ratio $\Delta H_c/\Delta H_g$*
Wood (fir)	4.0	0.435	9.2
Cellulose	4.0	0.418	9.6
Methyl alcohol	4.8	0.286	16.8
Polyurethane	6.2	0.364	17.0
Polyisocyanurate	6.3	0.364	17.3
Polyethylene	9.8	0.555	17.7
Polystyrene	9.1	0.420	21.7
Gasoline (octane)	11.4	0.14	81.4

*The higher the ratio, the more favorable are conditions for sustained combustion.

burning, but once it begins to burn it contributes strongly to sustaining the fire and to the total heat output of a fire. Materials at the two extremes join forces in most fires.

As tags on materials for household adornment frequently testify, insurance companies and fire-underwriter laboratories conduct intensive testing intended to rate the behavior of materials in a fire. This involves principally exposing them to sources of radiant heat, which can be more reliably standardized than actual fires. Tests have established two watts per square centimeter (cm^2) as the threshold at which many fuels will ignite. Upon ignition, the flux of heat at the surface is the sum of the heat from the external source and the heat radiated back from the material's own flame. The flows of heat into and out of a surface can be measured, as shown in the figure at right. Some of this energy goes into the decomposition of the sample, converting the solid into combustible gas, and is returned in the flow of energy and mass into the flame of the burning sample. The heat feedback from the sample's own flame helps importantly to sustain its burning, as measured by the numbers in the table on page 91.

In general, the fire testing of materials affirms commonsense judgments. Materials generally considered to be highly combustible feed back more energy than those not thought to be so. Thus, compared to gasoline, alcohol does not sustain luminous flames; its faint blue flame does not radiate strongly. It would contribute relatively low feedback to a fire.

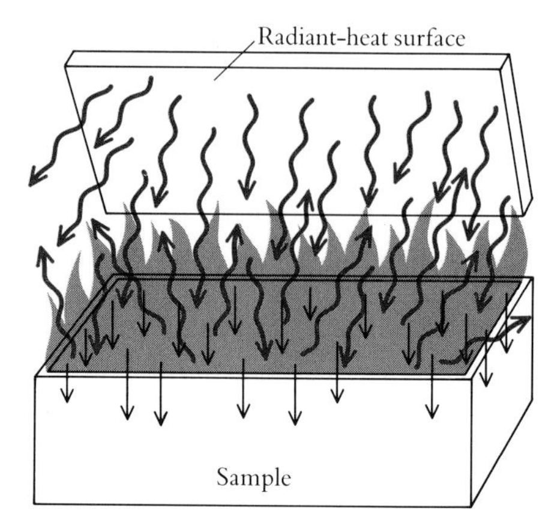

The flux of heat at a surface under an external source of radiation is diagrammed here. The largest inward flow is, of course, by radiation from the source above the surface, which has raised the temperature of the surface to the ignition point. As a result there is an additional flow of heat to the surface by radiation from the flames. From the surface some heat is reradiated away, and some heat moves by conduction into the material, balancing the flux.

Measurements of fire behavior must reckon, however, with many novel materials in furniture, draperies, and clothing that have come from the laboratories of the chemical industry. These have behaved in ways that have compelled revisions of testing procedures and standards.

In a test of urethane and isocyanurate foam plastics, the burning rate was found to increase with external heat flux, just as one would suspect (*see the figure below*). All the way to the test at four watts of radiation per square centimeter the material called AC lagged the procession. At that point, its burning rate increased sharply. If the tests had been run at only two watts per square centimeter, AC would have ranked at the top of the test, as fire "resistant." At four watts, its standing is close to the bottom.

Clearly something happens in this material at four watts per square centimeter. Most chemical reactions are strongly dependent on temperature, and the reaction rate, combustion reactions included, increases smoothly as a function of temperature. In fire tests, the rate of heat release is a good measure of the reaction rate, and the intensity of incident radiant energy from the external source correlates well with the material's temperature. When an abrupt change occurs, it signals that a new reaction has come into play or that, of several reactions involved, one is more temperature dependent than the others and has become dominant.

Heat Received from Flame (Back to Surface), q_F

Material	q_F (watts/cm^2)
Benzene	8.0
Styrene monomer	8.0
Polystyrene	8.3
Polyethylene	5.4
Polypropylene	5.0
Methylmethacrylate monomer	7.0
Polymethylmethacrylate	5.4
Polyurethane foams	7.0–8.0
Heptane	4.9
Ethyl and methyl alcohols	4.3
Wood (fir)	2.5
Phenolic polymer	2.3
Fire-retarded plywood	1.0

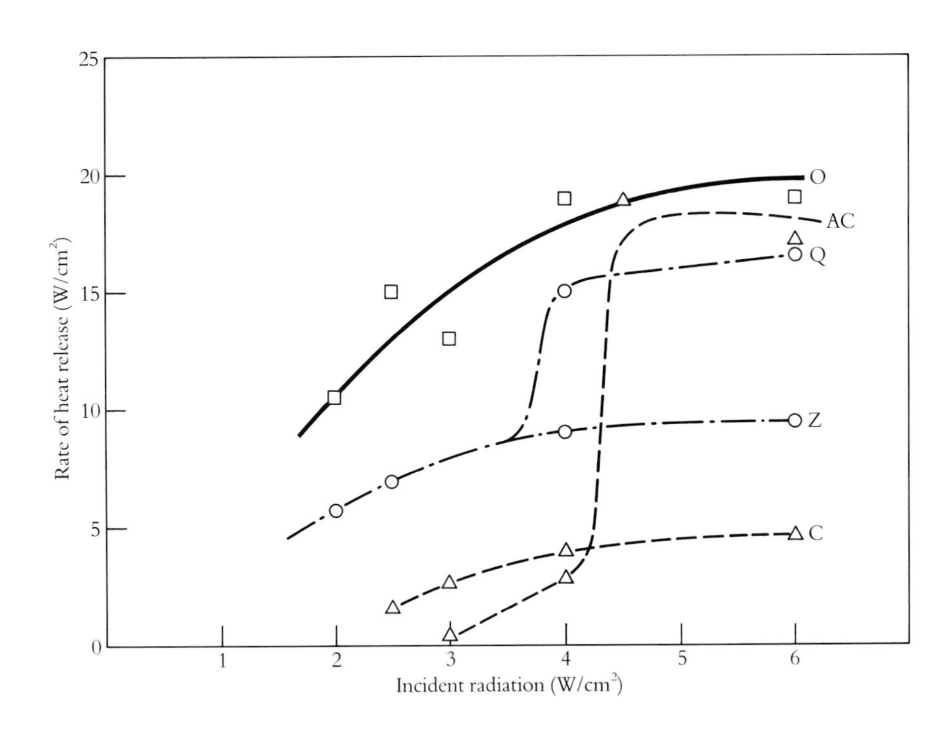

A comparison of the burning behavior of five different kinds of foamed plastic showed "AC" to be the most fire resistant until the flux of radiation in the test was boosted to four watts per square centimeter. Thereupon AC rated second-worst. The burning behavior of sample Q exhibited a similar temperature sensitivity at about three watts per square centimeter.

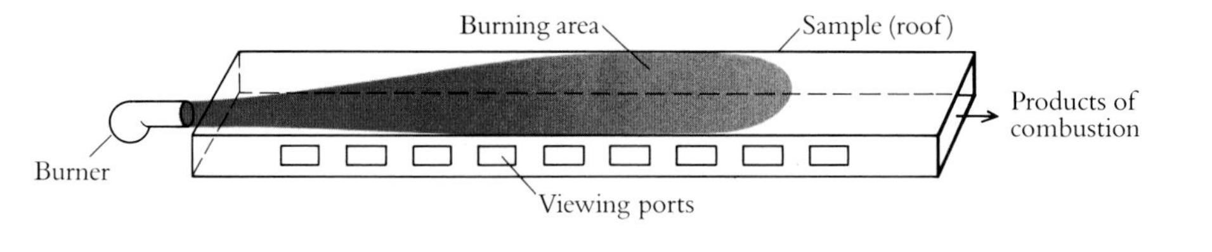

Tunnel test puts the material in question into place as the ceiling of a one-foot-high, 25-foot-long rectangular tunnel for exposure to a standard flame from a burner introduced at one end. The distance the flame propagates down the tunnel constitutes the measure of flammability. Ports in the side of the tunnel facilitate observation and measurement.

If a chemical fire retardant had been involved in the test of AC, one would suppose that the retardant had reached its temperature of decomposition. In this case, the sudden change in behavior had to be attributed to the material itself. Unraveling the competing mechanisms in burning materials has brought advanced techniques for the study of chemical kinetics into materials testing.

The purpose of fire tests is, of course, to predict how materials will behave in real fires. To design a controlled test invariably requires the fixing of a number of variables. By definition, that means the test represents only one or a few of the many combinations of conditions that may exist in a given real fire. A given test may expose only one side of a piece of flat material in one orientation for one length of time to a heat source of one intensity. The choice of conditions represents one, but clearly only one, possible real fire and may not even come close to representing the situation (say, six watts per square centimeter instead of two) to which the material may be exposed in a fire.

The designers of fire tests have called on considerable experimental ingenuity in their efforts to mimic aspects of real fires in the laboratory in order to make a rank ordering of materials. They developed some tests by adapting situations that recur in real fires.

Thus the "tunnel test" puts the material to be tested in place as the ceiling of a long rectangular passageway about one foot high and 1.5 feet wide. A large premixed flame is introduced into the tunnel at one end for a fixed period of time, and its propagation down the tunnel and the final distance the flame front has traveled, if indeed it stops before the end, are recorded. Combustible samples on the ceiling of the tunnel supply fuel to the burner flame and cause the fire to run farther down the tunnel. After correlation of the results of tests of materials with their observed performance in real fires, the tunnel test was established in 1950 as the standard for predicting the way flames would spread on the surfaces of all materials in real fires. Materials that came through the test with a response under a certain value got the rating "noncombustible."

It came as a rude shock when a variety of synthetic-polymer structural and decorative materials did not behave in real fires in accordance with their rating by the tunnel test. Severe fires have involved plastic-foam insulating materials that had received good ratings from the tunnel.

Factors in Considering Fire Behavior

Material properties	Density Total heat content Heat capacity Thermal conductivity Chemical analysis Heat of gasification
Behavior of samples in fire tests	Ease of ignition Rate of heat release Rate of surface flame spread Rate of smoke release Rate of toxic gas release Radiant power from flame
Critical variables in a fire in an occupancy	Temperature vs. time Smoke particulates vs. time Toxic gas vs. time
Impact on life	Thermal effects Toxic effects

Of the many factors to be considered in attempting to predict the behavior of materials in a fire (*see the table above*), only the physical properties of the materials can be reliably established by bench tests. These properties taken alone are not sufficient for predicting performance in real fires. The single exception is testing for ease of ignition.

The presently accepted test for the ease of ignition of carpet materials, for example, employs a pill-size ignition source that closely simulates the glowing coal of a cigarette. The test for ignition of a mattress requires the use of a lighted cigarette, a real fire test. Tests for other factors involve adjustable parameters on which agreement has not been easy to reach. Bench tests on isolated samples of material, it is reluctantly conceded, cannot provide data for other situations in which the configuration of the material, heat flux, air flow, and so on may be different from their representations in the test. There is a long distance, quantitative and qualitative, between the testing of a material with a standard source of heat and its sudden engulfment in a room full of roaring flames.

Because investigators learned they could not generalize from the results of laboratory bench tests of materials in isolation and could not reliably combine or integrate results from series of such tests, they turned to experiments with full-scale fires. Apart from those conducted at the National Bureau of Standards, in which I have been engaged, full-scale fire experiments have been conducted principally by Factory Mutual Research Corporation, Underwriters' Laboratories, Southwest Research Institute, SRI, Inc., the University of California at Berkeley, and Ohio State University.

Suppose that the problem is to evaluate various materials contending for use as cushioning on seats in subway cars. The investigator will try to obtain an actual car for use in tests. If that is impossible, a full-scale mockup will serve, preferably made of the same materials as the actual cars. The investigator then ignites a series of test fires, attempting to duplicate ignition sources and locations that are considered likely to exist in real subway operations. The same tests are applied to each of the cushion materials under consideration and, for comparison, to a car with unpadded seats. The temperature, smoke concentration, and toxic-gas concentration are measured at various places in the car as a function of the time after ignition. Limiting conditions must be established at which survival will become impossible, thus making it possible to estimate the time available for passengers to escape from the subway car after a fire is detected. The result will be a lengthy report with tables and charts for the use of the officials who will choose the materials for the seat cushions. They will, of course, consider many other factors such as price, future availability of replacements, and comfort in making their decision.

To fire-test a line of proposed subway cars, the National Bureau of Standards built a mockup consisting of a full-size wall, including a window, and ceiling panels, flooring material, and seats matching those to be used in the actual cars. Fires on or under the seats were started by lighting paper trash or lighter fluid (to simulate an arson attempt).

Full-scale testing of this kind is expensive, but its importance is being recognized ever more widely. When the BART subway system was built in the San Francisco–Oakland area in the 1960's, it was the first completely new subway system constructed in the United States in half a century. The designers set out to incorporate in the cars new design and materials that had not been used in subways before. The seat cushions and floors of the cars were made of a plastic material chosen in large part because it had proved highly fire resistant in bench tests, but it was tested without the benefit of the full-scale experimental procedures now established.

When a real fire occurred after a few years of operation, it reached temperatures far greater than those recorded in the tests. At these temperatures, the fire-resistant plastic melted and burned intensely, producing dense clouds of smoke and toxic gases. The passengers were able to escape onto another train, which carried them to safety, but one firefighter was killed before the blaze could be extinguished. The seat cushions were later replaced throughout the system with a different material, but the cars with the original plastic floors are still in use. A full-scale test under realistic conditions would have been far less expensive than the actual fire and its consequences.

From many full-scale tests came the now well-understood scenario of the fire in a room described at the beginning of this chapter. Research workers have recently tested aircraft fuselages, hotels, single- and multifamily dwellings, mobile homes, buses, subway-car mockups, mockups of navy ship interiors (including submarines), mockups of warehouses filled with combustibles, mockups of petroleum tanks, and so on and on. These experiments may

The burned-out interior of a subway car of the San Francisco Bay Area Rapid Transit (BART) recalls the fire in 1979 that betrayed deficiencies in the selection of materials for the fitting and furnishing of the cars. Reconstruction of the fire shows that the seat cushions proved especially vulnerable to the high temperatures generated after a high-voltage electrical short circuit. Fire in the seat cushions spread to the walls, ceiling, and floors, all of similarly novel material and construction. Casualties, including one fatality, were caused by smoke inhalation. Fire fighters, wearing gas masks, were blinded by smoke so thick "it looked like black ink."

cost as much as $100,000 each. The results of such testing have been incorporated in the designing, equipping, and furnishing of the kinds of people-containing spaces tested. Those results have also taught caution to the officials who are responsible for such decisions about the safety of the public.

The full-scale tests have also shown a way to the development of a conceptual approach to the unwanted fire that permits testing of hypotheses on a smaller scale. Use of a smaller scale is the most obvious solution to the high cost of full-scale tests. A set of scaling rules for testing at the one-fourth scale has been validated at the National Bureau of Standards for the evaluation of interior linings and some furnishings of dwellings. Not every parameter behaves acceptably when scaled down. Flame height, for example, does not scale down in the same proportion as other parameters. The result is that the fire conditions in the small-scale test measure out as less severe than in full-scale tests. The difference is not great, and corrections can be made for it.

From bench tests, full-scale and now smaller-scale fire tests, have come enough data and insights to make possible the building of mathematical models. These may tell us more than we could learn otherwise about what goes on in real fires. Early attempts to model a fire in a small room set necessarily modest goals. A successful model helped to establish the crucial role that the door or the window opening plays in the time course of a fire.

That model treats the gases in the room as if they are fully mixed and equilibrated. The solids in the walls, ceilings, and floor are likewise treated as receiving uniform thermal exposure and, for the purpose of the exercise, as noncombustible. Access to air outside is through a single door. With fire on a single fuel in the center of the room, its course can be tracked in terms of a single gas temperature as a function of time. Even with these simplifying assumptions, the situation presents formidable conceptual complexity. Two equations—the ideal gas law, which relates temperature, pressure, and density, and the equation relating the pressure, density, and velocity of a flowing fluid—plus equations for heat transport by conduction, convection, and radiation are all required to manage the heat exchange between the hot gas accumulating inside the room and outside, an exchange in which heat loss through the walls and other room boundaries plays an important part.

The energy released and the fuel consumed during the first time unit are calculated from the initial burning rate. The effects of the released energy on the temperature and so the other properties of the air and on the temperature of the room boundaries, and the loss of energy through the boundaries and room openings are computed. With the results of this computation as the new starting values the equations are computed for the next time interval. Iteration of these computations again and again establishes the course of the fire on a time-temperature curve.

The flow of air into and out of the room is driven by the buoyant forces of convection. The rate of air flow and hence the burning rate vary as a function of the doorway height and area until they are so large the fire might as well be burning outdoors; that is, there is no limiting effect. Two curves from a study conducted in Japan in the early 1960's show this dependence. A small opening may starve a fire for oxygen; increasing the opening increases the rate of burning. The interesting dependence is on door height, with mass flow varying as the height of the doorway raised to the $\frac{3}{2}$ power. Experiments verify this behavior.

The next generation of mathematical models is more complex, subdividing the volume of air in the room into smaller subvolumes. The course or the effects of the fire in each subvolume are calculated independently, and the effect of the change in each subvolume on its neighboring subvolumes is calculated. Even retaining the simplifying assumptions of incombustible room surfaces, a single item of fuel, and uniform surface conditions within subvolumes, this increase in model complexity produces a huge increase in the number of calculations to be performed at each iteration. Many subiterations are needed to arrive at an approximation of the interactions among the several subvolumes in the room. The results of these studies show that conditions in the air are far from uniform throughout the room. Not long after the fire begins, the upper part of the room becomes filled with hot, dense smoke and exhaust gases. The hot gases and soot particles radiate significant amounts of heat. The geometry of the room surfaces and the nature, shape, and size of the hot, radiating layer play an important role in determining the amount of energy feedback to the flame zone, and hence the burning rate of the fire. With such a model on the computer, a number of conditions can be changed and the sensitivity of the outcome to such changes can be observed.

The results generated by the model were not unexpected; what the model showed had already been observed in studies of real fires. Higher wall conductivities help to remove heat from the fire room, decreasing radiative feedback and reducing fire severity. An increased concentration of soot particles in the hot, upper part of the room increased radiative feedback and fire severity, so that materials producing extensive smoke as they burn can be dangerous in more ways than in their direct threat to human lives, through reduced visibility and increased lung damage. Although these studies did not reveal important new information about fire behavior, the improved correlation between the model and observations of actual fires suggested that still more detailed models would be practical substitutes for full-scale testing; so work in this area continues.

Such detailed models have been developed for more limited aspects of a room fire. One model studies the spread of fire up a wall covered with a combustible material. A source of ignition heat (such as might be provided by an electrical short circuit) is assumed at the base of the wall, and the rising plume of hot gases from that heat source is calculated. Radiation from the plume, heating of the wall, and reradiation to the plume are all estimated, and the calculation is then iterated for the next time interval. The results from this model have proved to correlate closely with the results of certain large-scale tests.

Current complete-room models take into account room dimensions and geometry, the heat of gasification and heat of combustion of the fuel, pyrolysis temperature and heat capacity of the fuel, flame temperature, the thermal properties of room surfaces, the rate of mass loss from the fuel, the thermal conductivities of the walls, the size and shape of doors and windows, and so on. The introduction of supercomputers in the mid-1970's provided computational power to accommodate these more detailed and refined models. The model now in use at the National Bureau of Standards for the study of buoyant flow inside a fire room is able to compute details of the turbulent flow of gases and even to keep track of the changing positions and sizes of soot particles. The room volume is divided into some 70,000 subvolumes (or cells) and as many as 12,000 representative particles can be tracked in their travel through these volumes. The equations involve five dependent variables that are followed through 3,000 time steps, or major iterations. Some 400,000 equations

Fire Modeling

In Galileo's words, "Mathematics is the language of nature." A mathematical model of a natural phenomenon sets its variables out in their relation to one another. Thus, the perfect gas law reads:

$$PV = nRT$$

where, for a fixed amount, n, of gas, P is pressure, V is volume, R is a constant, and T is temperature (in kelvins).

The equation tells a great deal about our model at a glance and, simple and approximate though it is, about the behavior of a gas. It shows the direct relation between pressure and temperature at constant volume and between volume and temperature at constant pressure, and the reciprocal relation between pressure and volume at constant temperature. Observed relations among the three variables in a gas were fully described in dimensionless terms, first written for pressure and volume, by Robert Boyle in 1662, and later for all three by Jacques Charles in 1782 and Joseph Gay-Lussac in 1802. Closer observation brought correction of the equation in the 19th century to account for the molecular forces of repulsion at high pressure and for the volume occupied by the constituent molecules of a gas.

The perfect gas law in its original form suffices, without correction, for our purposes in attempting to construct mathematical models of the chaos of an unwanted fire. For this complicated situation, with its many variables, we set up equations that correspond to our observations or to our deductions and assumptions about what we cannot observe directly. In three differential equations that give the rate of change of density, motion, and energy we make the most general statements about the system in terms of changes of sets of physical quantities that encompass all the physical variables. Thus the energy equation reads:

<table>
<tr><td>rate of gain of energy =</td><td>rate of gain
by convection</td><td>+</td><td>rate of gain
by conduction</td><td>+</td><td>rate of gain
by radiation</td><td></td></tr>
<tr><td></td><td>+ rate of work
done against
gravity</td><td>+</td><td>rate of work
done against
pressure forces</td><td>+</td><td>rate of work
done against
viscous forces</td><td>+ · · ·</td></tr>
</table>

For each term on the right we need a model law governing the behavior of that variable. In the first term, for convection, we use the perfect gas law to give the simple relation between pressure, volume, and temperature needed in approximating the flow of the convecting gases. In the second term, for conduction, we use Fourier's law; for gravity, Newton's law; for viscous effect, the simplest model of an ideal, or Newtonian, fluid, and so on.

We insert the numbers for the starting conditions in the equations and proceed to run them in one, two, or three dimensions on a computer, using special techniques to expedite the numerous calculations. The design and operation of such a model requires the collaboration of a mechanical engineer or physicist, a mathematician, and a programmer skilled in scientific computing.

are solved in each iteration. Over one billion items of data are generated in a single run of this model, of which only a tiny fraction may be used by another computer program to generate a graphic display that summarizes the results for study by the investigators.

In the present state of the modeling art, it is possible to predict how long it will take after ignition for a fire to develop that completely involves all of the fuel sources in a room. For certain geometries that have been studied intensively in full-scale fires, the models now perform with an accuracy of plus or minus 10 percent. One test, setting a fire in a furnished bedroom, burned seven minutes to flashover. That time was predicted to within one minute by a prior run of a model.

These studies bear on crucial decisions that confront designers and fire-safety officials. Does a given design of a room or building give sufficient time for escape? If it is judged that 10 minutes are required, then a model that can predict the time to plus or minus one minute comes close to being adequate. Mathematical models for use by designers, regulatory officials, and materials manufacturers are nearly at hand. Software for such models should see application within the decade of the 1980's. This will produce a great change in the way building and fire codes are prepared and used. Codes have been changed after major disasters in the past, as will be seen in the next chapter, but rarely completely overhauled. Computer models will compel such overhaul.

More than half of all fire deaths are caused, however, not by burns—even those sustained in the catastrophic flashover—but by inhalation of the hot combustion gases that flood into nearby spaces from the burning room. A study of recent figures in the state of Maryland shows that 80 percent of all deaths occur within six hours of the fire, the other 20 percent dying half from burns and half from chemical pneumonia as a consequence of inhaling the

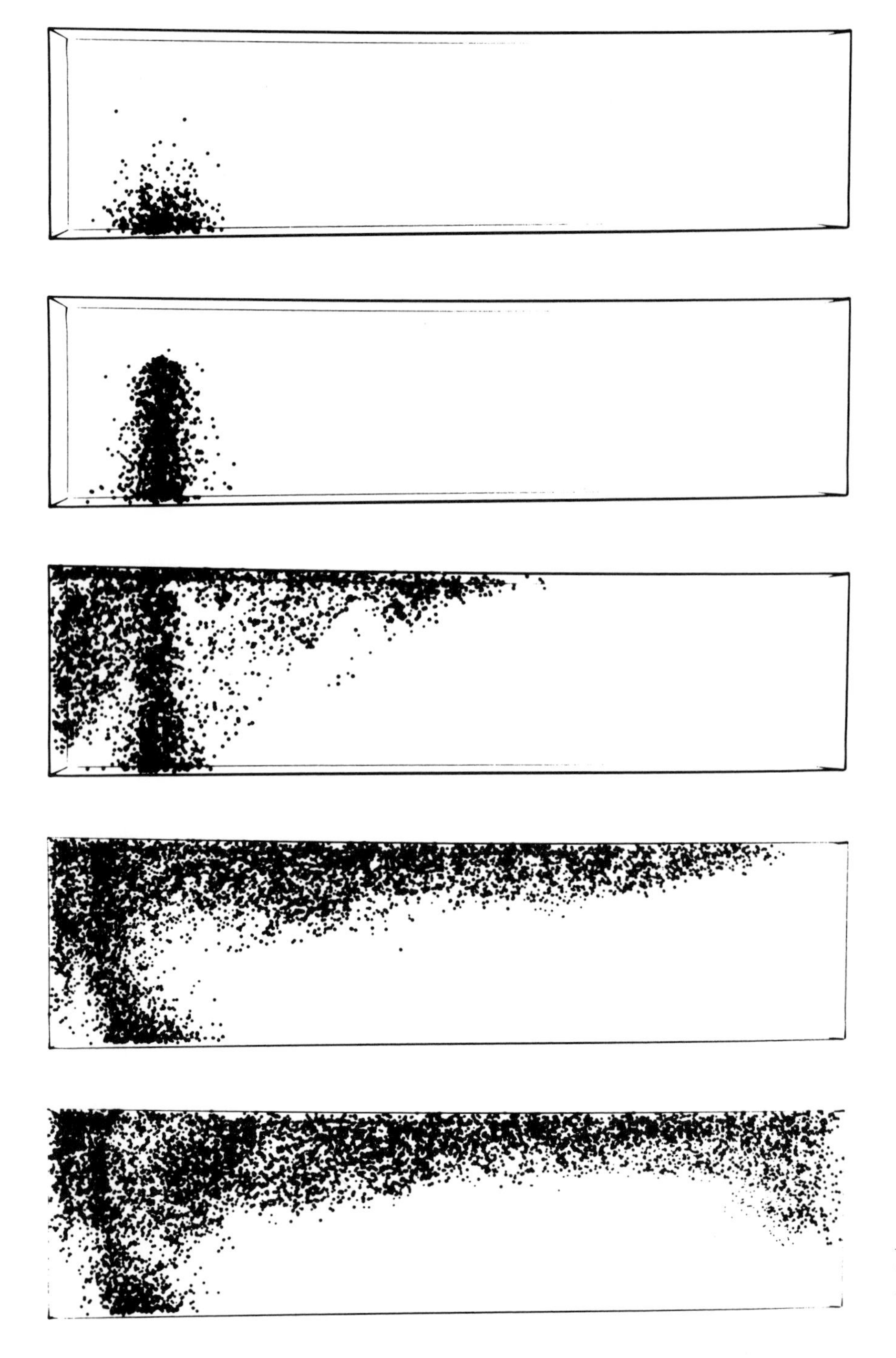

A computer simulation of fire in a room shows the development of the fire plume and then of the hot gas layer. This model experiment was run by the National Bureau of Standards on a supercomputer, which solved the large number of equations necessary to establish the changing conditions in the total volume of the room, subdivided for the purposes of calculation, and to track the release and dispersal of 70,000 particles, which served as a means of following the gas motions.

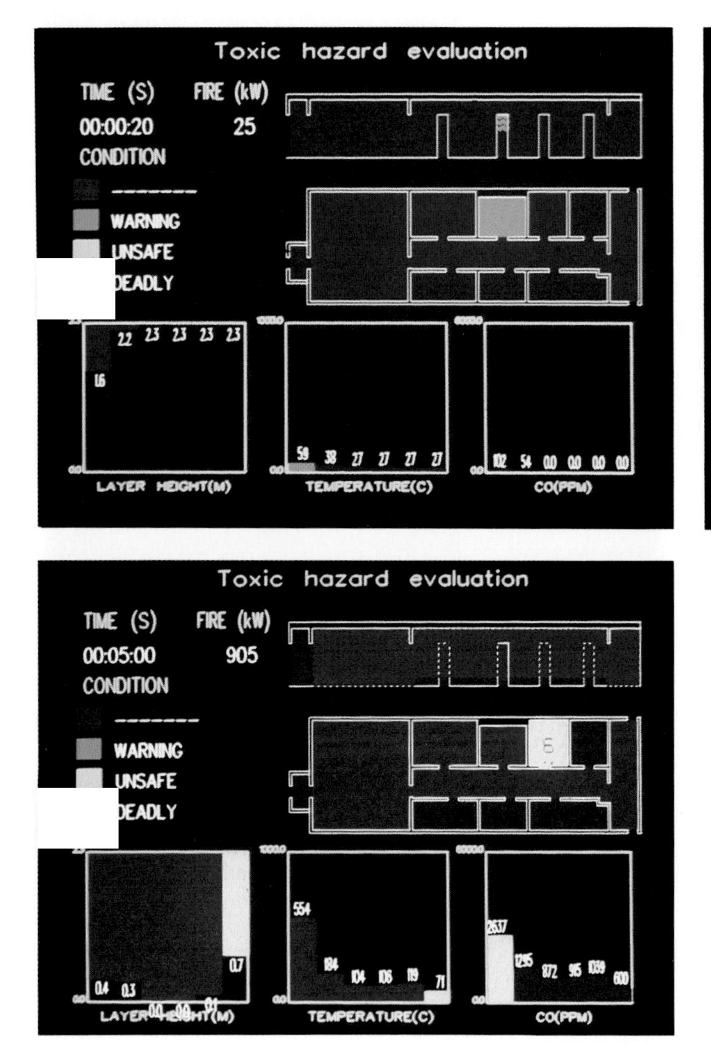

Movement of hot smoke and combustion gases is modeled on a computer at the National Bureau of Standards to observe their spread through a suite of rooms from a fire in a single room. A cross section of the corridor connecting the rooms is at the top; a plan view of the suite is below. The intensity of the fire is monitored in kilowatts (kW), and the progressive increase in the lethality of the temperature and the carbon monoxide concentration in the deepening gas layer is tracked room by room in the bar graphs at the bottom. The door to the room at the right of the fire room is set slightly ajar; the doors to the next room and the rooms across the corridor are shut. At five minutes, the layer of hot gas has reached the floor in the corridor and connecting rooms and its temperature exceeds the boiling point of water.

combustion gases. About 60 percent of the fire deaths occurring in the first six hours were caused by the toxic effects of the combustion gases, of which carbon monoxide is the major component and principal killer. In the deaths of another 20 percent preexisting circulatory disease made the victims especially vulnerable to carbon monoxide poisoning. The other 20 percent died of injuries and heart attacks attending their experience of the fires. In one-third of all cases, incidentally, high levels of alcohol were found in the victims' blood, and one-half of the victims age 30 to 59 years had an amount of alcohol in their blood above the legal intoxication level.

An experimental fire in a mockup warehouse gets violently under way. With dense stores of materials, almost invariably flammable and most often with high heat content, warehouse fires burn very hot and for a long time. The spaces between the shelving in this experimental setup supply ample draft space to bring air into the fire burning at each level in the high stacks. Sprinklers at ceiling level cannot reach these spaces and must be supplemented by sprinklers set into the racks.

Especially from an unwanted fire, the combustion products are not those predicted by the equations for complete combustion. Combustion is not complete. If it were, fuel containing hydrogen, carbon, and oxygen would yield only carbon dioxide and water:

$$C_xH_yO_z + \underset{\text{(excess)}}{O_2} + \text{heat} \rightarrow xCO_2 + \frac{y}{2}H_2O$$

Complete burning is hard to achieve even in a controlled furnace. This is partly because it is difficult to secure uniform mixing throughout the combustion zone and because some particles are too large to burn completely in the short time they reside in the flame, producing a sooty flame. There are regions high in fuel and low in oxygen and regions low in fuel and high in oxygen. The fuel-rich regions yield gases that survive the combustion zone and emerge only partly oxidized.

Incomplete combustion produces carbon monoxide as its principal product. In itself carbon monoxide is sufficiently lethal. Other poisonous gases, such as hydrogen cyanide, are identified in fire victims at autopsy, but never in the absence of carbon monoxide.

The increasing volume and variety of synthetic polymers that burn in unwanted fires raise new questions about the hazards from toxic effects. Although all polymers produce carbon monoxide—in the case of nylon and wool, 30 to 60 times their output of hydrogen cyanide—other poisonous gases must be reckoned with. Certain polymers produce additional gases that are toxic at low levels. Thus polyvinyl chloride yields hydrogen chloride gas, among other gases. Hydrogen chloride is 10 times more toxic to humans than carbon monoxide. Hydrogen cyanide, from nitrogen-containing polymers, is about as toxic as hydrogen chloride. Since wool and silk as well as synthetic polymers such as nylon and acrylonitrile contain nitrogen, hydrogen cyanide is frequently encountered in building fires. Quite apart from the hazards posed by new materials, burning wood produces a complex mixture of combustion gases besides carbon monoxide. Among them are aldehydes that produce eye irritation and copious tearing; this reaction inhibits escape by interfering with vision.

Careful analysis of gases from experimental fires reveals dozens of additional compounds; some of these may be extremely toxic even at very low levels. Many of the substances have not been evaluated by toxicologists and are not to be found in the handbooks of toxicology data. To compound the problem there are few data on mixtures of gases. The data that have been obtained on mixtures do not support the assumption that the toxicity of a mixture is simply the sum of the toxicities of the components evaluated separately. It is not enough, therefore, to compute the toxicity of the combustion products of a new polymer from a simple chemical analysis of the fire gases.

Toxic Effects of Gaseous Products from Heating Various Materials

	Concentration for 50 percent mortality (mg/l)*	
	Nonflaming heating	Flaming heating
Polyurethane foam, flexible	14	43
Douglas fir	19	24
Polyurethane foam, rigid, with fire retardant	>65	14
Polyurethane foam, rigid	55	17
Polystyrene foam	66	29
Polystyrene foam with fire retardant	>65	36

*Rats were exposed for 30 minutes and then observed for 14 days. The numbers are for one particular kind of material in each case and are not representative of an entire class.

If not chemical analysis, then what? The answer is bioassay, the measurement of the response of test organisms to combustion products. The choice of organism is a compromise: the lower the life form, the more economical the test procedures but the less certain the extrapolation to humans. The rat or the mouse is usually chosen because they are mammals but at the low end of that family.

There is, as yet, no accepted standard against which to measure the toxicity of combustion products. A reference point for comparing unknown materials is needed if only to enable the toxicologist and the fire researcher to talk to each other. Some maintain that a material from which the combustion products appear to be no more toxic than those from wood—say, from Douglas fir—should be acceptable. Some feel that wood smoke is too toxic to serve as a reference.

Present toxicity testing relying on animals is of such uncertainty that differences of a factor of 2 or less are not statistically significant (*see the table above*). The animals are exposed to measured concentrations of the combustion products of the material in question. The point at which half the test animals die establishes the lethal concentration, designated LC_{50}. In general, the effects are different for flaming and nonflaming modes, for, as one might expect, there is much more complete combustion in the flaming mode.

Wood smoke is a fairly noxious material, containing as it does both gases that are toxic at moderate levels and strong irritants to the eye and throat. Hence saying that the products of combustion of a substance are no more toxic than wood smoke is hardly an endorsement. The point is that wood products are ubiquitous in the built environment, and society would have little stomach for banning or restricting their use. Thus even though the combustion products wood smoke carries have severe effects on organisms, it will undoubtedly serve as a basis for making at least screening-level decisions. If one has developed a new material and wishes to see if its combustion products pose an unusual hazard, it seems reasonable to have a toxicologist conduct studies and compare the results with those for wood. If the substance appears to be in the same class as wood, that result would not be reason to stop developmental work. If the results were substantially poorer than those for wood, on the other hand, more detailed toxicological studies would be warranted.

Ultimately the evaluation of the hazard of any new material must go beyond the bench screening test that now provides information from the thermal decomposition of a test sample in a limited number of ways in one geometry. The amount of the material proposed for use in a given room, its location in the room, the other materials present, the resistance of the material to thermal effects, the occupancy type, and the most likely fire outbreak must all be reckoned with. The material may be resistant to thermal decomposition or it may be a nearly perfect heat reflector in its form for the intended use. These factors, too, must be taken into account. Work on problems in this concatenation is just getting under way.

The Maryland fire-fatality figures underline the urgent need for a better understanding of the smoke hazard. Hot gases in the smoke take casualties at a distance from the fire itself. In a sensational fire at Yonkers, New York, in 1965 the hot carbon monoxide in black smoke that erupted from the explosive burning of 1,000 pounds of polymethyl methacrylate screening used to adorn a public auditorium killed 12 people; the building was otherwise largely undamaged by the fire. The victims were not in the auditorium but in a room two floors up and on a connecting corridor; people in other rooms on the same corridor who had the doors closed went unscathed.

It is most often the same heat and incomplete products of combustion in the smoke that spread the fire from the room where it starts. Full-scale test fires combining a room and a corridor and a room, a corridor, and another room have demonstrated the scenario for this process. At the flashover of the fire in the room, smoke and tongues of flame erupt through the door into the corridor. The smoke and hot gases displace the colder air at the ceiling of the corridor and flood down its length. Cool air entering through any other opening in the corridor floods along the floor toward the fire room. Between the hot gases flowing away from the room at the ceiling and the air flowing toward it at the floor there is for a while a neutral region of no flow near the middle. Immediately outside the doorway the temperature at the corridor ceiling reaches 550 to 700°C, and the sooty smoke heats the ceiling and the upper walls of the corridor. The heat flux at the floor often does not reach the ignition level of about two watts per square centimeter so long as there is no fuel in

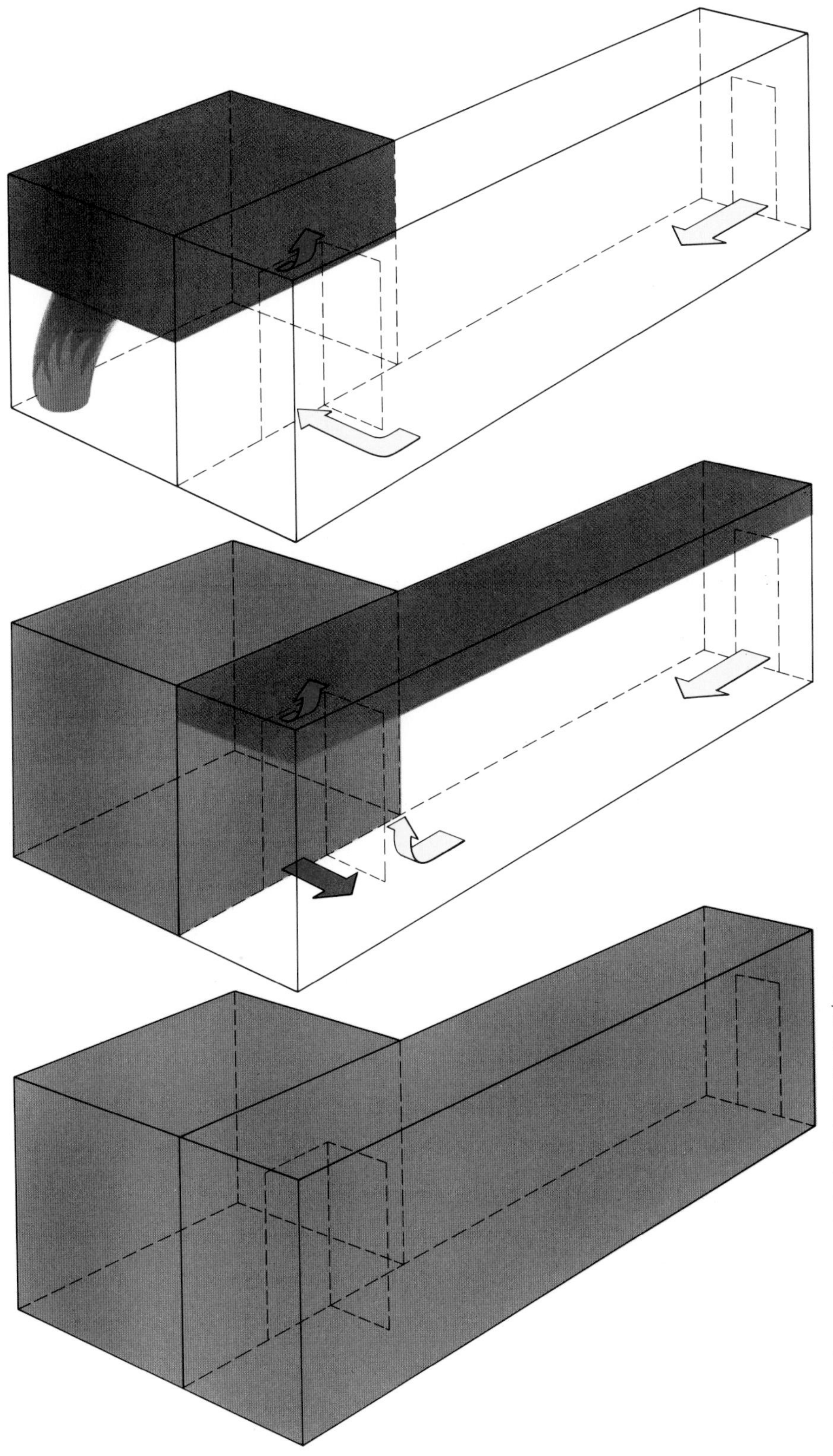

An experimental fire in a room-corridor configuration shows how fire spreads from the room of origin to the other parts of the building. With a rapidly growing fire in a room, flames and soot-laden smoke erupt through the door into the corridor, starting a buildup of a layer of hot combustion gases under the corridor ceiling. Air running along the floor of the corridor supplies oxygen to the fire, now in vigorous steady-state combustion in the room. After a few minutes the fire advances through the door onto the corridor floor; convection of combustion gases at the door now blocks the flow of air along the floor into the fire room. Oxygen-starved, fuel-rich gases build up under the ceiling of the room for several more minutes. When they spill into the corridor and encounter oxygen, they burn in an explosive flash-over.

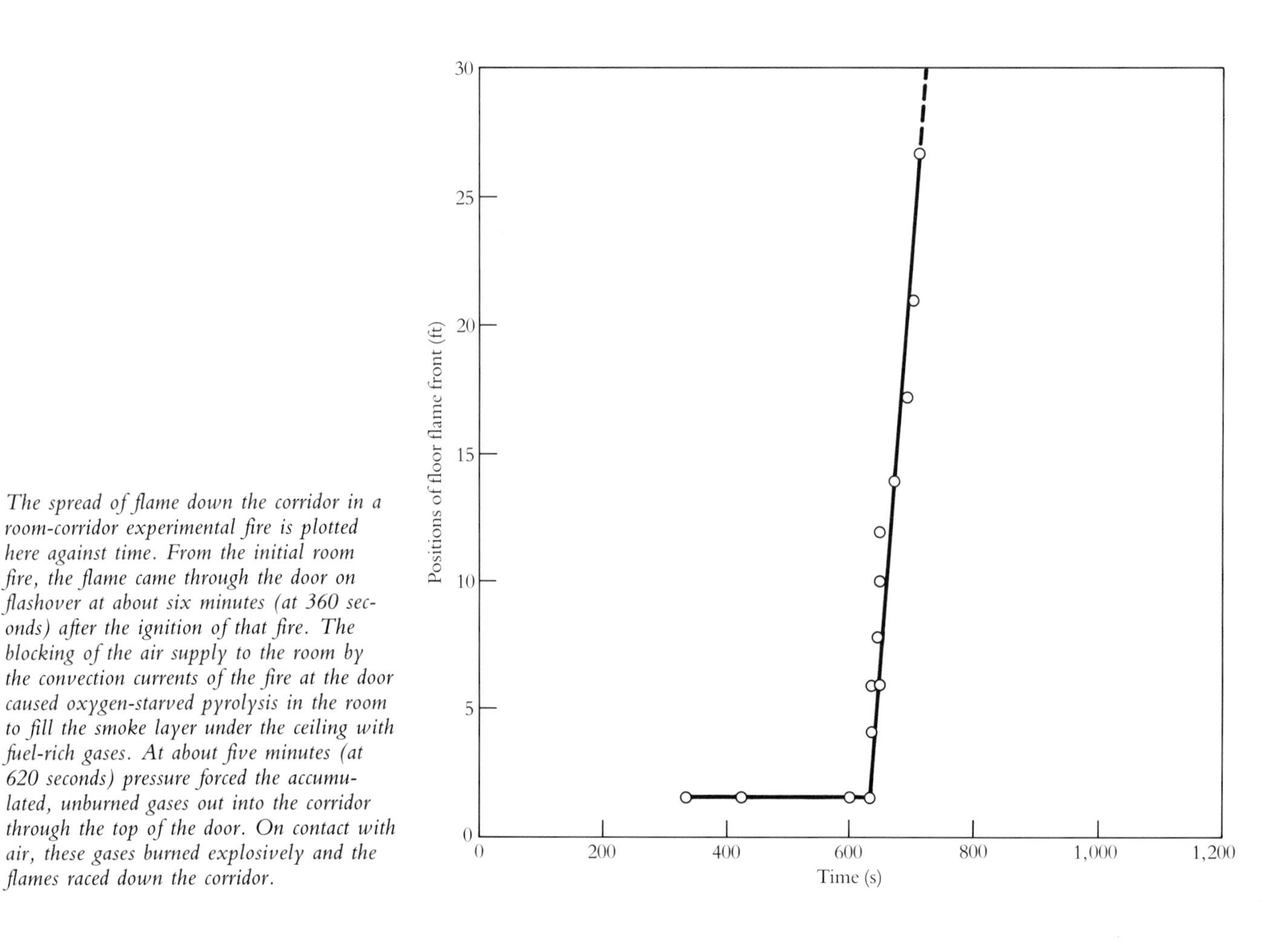

The spread of flame down the corridor in a room-corridor experimental fire is plotted here against time. From the initial room fire, the flame came through the door on flashover at about six minutes (at 360 seconds) after the ignition of that fire. The blocking of the air supply to the room by the convection currents of the fire at the door caused oxygen-starved pyrolysis in the room to fill the smoke layer under the ceiling with fuel-rich gases. At about five minutes (at 620 seconds) pressure forced the accumulated, unburned gases out into the corridor through the top of the door. On contact with air, these gases burned explosively and the flames raced down the corridor.

the corridor to intensify the fire. Flames on the floor of the fire room begin to emerge, however, through the door onto the corridor floor.

If the materials of construction in the corridor are combustible, then, as in the fire in the room, the situation in the corridor proceeds toward a further abrupt change: a flashoverlike end point. In the experiment charted in the figure on page 105, the fire moved onto the corridor floor (made of red oak) at just over 300 seconds from the start of the fire in the room. It did not progress down the corridor for almost another 300 seconds. Then in about 60 seconds it traveled the length of the corridor. The last 300 seconds is the life-and-death decision time for fire fighters or occupants.

The corridor experiments have shown that this particular sudden spread of flame is caused, paradoxically, by depletion of oxygen. The fire on the floor of the corridor at the door to the fire room heats the incoming air and diverts it from the doorway upward to the corridor ceiling, where it joins the gases from the fire moving back down the corridor. Soon the inflowing air is completely blocked from the fire. When this happens, oxidation in the fire room stops. The unburned fuel in the room remains, however, at pyrolysis temperature and gasification of the fuel continues.

Soon these fuel-rich hot gases are flooding under the lintel of the door out under the ceiling of the corridor. There they encounter oxygen. Their sudden ignition sets the whole corridor aflame in a flashover like that in the room.

This flashover differs, however, from the one that engulfed the room. There it was the radiant heating of all the combustible materials in the room simultaneously to the ignition point. In the corridor it is the sudden ignition of hot, fuel-rich gases at the ceiling. The buoyancy of the hot gases thus carries the fire from room to corridor and thence wherever they can penetrate. If there is an open stairwell, the flammable gases will quickly rise to floors above, flaring into flame as they move. This kind of flashover may occur over very great distances. The common report from major fire disasters is that suddenly fire was everywhere. The explanation is undoubtedly the sudden ignition of large clouds of unburned fuel gases escaping from an oxygen-starved fire behind them.

Conflagration in the 31-story Andraus Building, São Paulo, Brazil.

6 *Fire!*

To anyone who has experienced the calamity we have been calling the unwanted fire, it is difficult to prove there has been progress in the prevention, control, and fighting of fire. A long perspective is required to make that progress visible. In a sufficiently long historical perspective, however, one can see in this country advances along three lines accomplished in prolonged but overlapping periods.

At the very least, in the first place, it can be said that a significant risk of exposure to conflagration is no longer a condition of urban existence. That was a risk built in to the first cities on this continent. The techniques of fighting, control, and prevention, learned in that order, of fires that would burn out whole sections of cities if not entire small cities were devised over the course of two centuries, from the early 18th into the first half of the 20th century. This advance was won by hindsight after painful and expensive lessons that had to be taught more than once.

The second advance might conveniently be dated from the fire that brought down the celebrated Crystal Palace in New York City in 1858 to the fire in 1926 which the second Equitable Life building in that city survived. This period, with its numerous hard lessons, brought the fire-resistant building into being. Within buildings, the third line of advance is bringing the containment of fire in the combustible contents and the defense of occupants against the fire and its lethal smoke.

If corresponding progress had been made on a fourth salient, against fire in the American home (to which the last chapter is addressed), our country would have a higher ranking on this measure of civilization.

It was Benjamin Franklin, of course, who organized the first fire-fighting company in America in the 1730's. In Philadelphia, as in other cities, the houses and shops, joined by party walls or separated by narrow alleys, were a menace to one another. Often they were there before central services, such as water mains, were laid in. The formally organized volunteer company was an advance over the ad hoc bucket brigade organized on the spot. A significant center of neighborhood social life, it raised funds for such equipment as lad-

Towering Inferno

It should be more widely appreciated that the invention and the success of the skyscraper—the high-rise building—stem from the writing and enforcement of the building codes that specify the materials and methods of their construction and crucial elements in their design. What can happen in the absence of such safeguards had an appalling demonstration in the fire that devastated the 31-story Andraus Building in São Paulo, Brazil, on February 24, 1972.

The fire began in a storage space filled with combustibles. It spread over combustible ceiling tiles, moved up open stairwells, and exploded through the windows of upper floors. Once it had broken out of the building, it traveled upward from floor to floor through closed windows by radiant ignition of combustibles inside each floor. The one enclosed stairway extending from street level to roof was not fitted with fire doors onto each floor, and the fire broke through from the fire floors into this single escape path, shutting it down. There were no automatic detection or extinguishing systems. The result is plain enough in the picture of the fire shown at the beginning of the chapter. Incredibly only 16 people died and only 375 of the 1,000 people thought to have been in the building suffered injuries.

In the United States, building regulations would prevent such a holocaust from occurring. Each year, however, we have serious fires inside large buildings. They have not developed to the extent of the São Paulo one because the regulations, even if imperfectly implemented, are stringent enough to prevent spread over the entire structure.

ders, pumps, and hoses in better preparation for the next emergency. In the absence of mains and hydrants the volunteers pumped water from a river or lake, sometimes with the companies pumping in a chain and competing to pump more water into the next company's reservoir than its crew could pump out, a triumph known as "washing." In Philadelphia the first hose was hooked up to a hydrant in 1803. Even in the biggest cities, fire fighting remained a volunteer enterprise until the second half of the 19th century. Philadelphia did not centralize its fire companies in a fire department until 1855 and did not put its firemen on the municipal payroll until 1870.

Major U.S. Fires

City	Date	Buildings lost	Lives lost
New York	1835	700	–
St. Louis	1849	400	–
Chicago	1871	17,500	766
Peshtigo, Wis.	1871	All	800
Boston	1872	776	–
Jacksonville, Fla.	1901	1,700	–
Paterson, N.J.	1902	525	–
Baltimore	1904	2,500	0
San Francisco	1905	28,000	1,188
Chelsea, Mass.	1908	3,500	18

The cities were thus not prepared for the catastrophes that overtook them when the season was dry or the winds were strong, and they fared little better in other seasons. In December, 1835, a fire raged out of control for three days in downtown Manhattan, the commercial heart of New York City. Fire companies came from as far away as Philadelphia. Freezing weather kept the city's hydrants and the pumping gear out of action. Rows of houses had to be leveled by blasting to make fire breaks. When the fire at last waned, it had destroyed 700 buildings.

The Great Chicago Fire in October, 1871, betrayed the absence of zoning and of the regulation of building construction in that city's anarchic growth. Records show the fire did start in the O'Leary family's barn, giving that much support to the legend that Mrs. O'Leary's cow had kicked over a kerosene lantern at milking time. Residences, shops, warehouses, factory buildings, 17,500 structures built mostly of wood, crowded together on 2,000 acres, burned to the ground in 24 hours. Nearly 300 people died. Fewer than 200 firemen were on hand to fight the fire, which was driven by a brisk, dry wind and ended by a heavy rain.

By 1872, Boston had three-inch water mains, hydrants, and 13 pieces of wheeled fire apparatus, including pump engines that could lift a hose stream three stories high. The great fire of that year burned 776 buildings, many of them six stories high, in an area about a mile square on the waterfront side of

The New York City fire of 1835 destroyed the commercial heart of the city, adorned by the Merchant's Exchange, shown here in its last hour. Stone facades belied interiors of wood. Fire hoses and hydrants froze in the cold weather.

Beacon Hill. The city put its fire department under a board of fire commissioners, built more firehouses, and installed bigger mains and more hydrants.

A big fire in Baltimore in the winter of 1904 brought fire companies from Washington, D.C., Wilmington, and Philadelphia, until 1,700 men were fighting the blaze. Because their hose couplings would not fit the Baltimore hydrants they had to pump water from the waterfront or haul it from there in barrels. By the time the wind, blowing at 25 to 30 miles per hour, had driven the fire to a natural firebreak at Jones Falls canal, it had destroyed 2,500 buildings on 25 acres of land.

Thereafter, the newly formed National Bureau of Standards (NBS) published a standard for fire-hose coupling threads and made available a standard set of dies for calibrating manufacturing devices. Such is the rate of amortization of municipal capital, however, that suburban hose companies must still carry adapters in order to connect their hoses to hydrants in downtown Washington. The NBS standardization, of course, makes this connection possible.

Across the mouth of the Charles River from Boston in Chelsea in 1908 one of the last great conflagrations in the United States killed 18 people and

The Boston fire of 1872 destroyed the on-shore installations of the city's prosperous ocean shipping industry. The city had outgrown its fire-fighting technology, with six-story buildings out of reach of the three-story pressure in the water mains.

burned 3,500 buildings exposed by their design and construction to the hazards, variously, of combustible roofing, unprotected structural steel, interior wood-frame construction, and proximity to long, low wood waterfront structures improperly storing combustible materials, including some liquid fuel. The 21 firemen and four pieces of apparatus funded by this impoverished municipal dependency of Boston were overwhelmed by the task, but neither could reinforcements from the rebuilt and expanded Boston fire department contain the wind-driven fire.

The fire-fighting services of most of the communities in the United States continue to be manned by volunteers; the local tax base now usually funds at least some of the capital equipment. Most of the population lives, however, in cities with fire departments manned by full-time professional civil servants. Everywhere, as in Boston, conflagrations and big fires brought funding of these departments on a scale more commensurate with need. Fire apparatus became more sophisticated and matched to the height and type of structure to be protected. City water departments installed larger mains, most being six inches in diameter and pressured to 50 pounds per square inch. Hydrants are everywhere installed within hose reach.

The Union Fire Company

In his *Autobiography,* Benjamin Franklin gives this account of the founding of the first fire company in America in 1736:

A horse-drawn, steam-powered pumping machine could lift water from a hydrant to upper stories in a fire.

About this time I wrote a paper (first to be read in the Junto, but it was afterward published) on the different accidents and carelessnesses by which houses were set on fire, with cautions against them and means proposed of avoiding them. This was spoken of as a useful piece, and gave rise to a project, which soon followed it, of forming a company for the more ready extinguishing of fires and mutual assistance in removing and securing of goods when in danger. Associates in this scheme were presently found amounting to thirty. Our articles of agreement obliged every member to keep always in good order and fit for use a certain number of leathern buckets, with strong bags and baskets (for packing and transporting of goods), which were to be brought to every fire, and we agreed, about once a month, to spend a social evening together, in discoursing and communicating such ideas as occurred to us upon the subject of fires as might be useful in our conduct on such occasions.

> The utility of this institution soon appeared, and many more desiring to be admitted than we thought convenient for one company they were advised to form another, which was accordingly done, and thus went on one new company after another till they became so numerous as to include most of the inhabitants who were men of property, and now at the time of my writing this, though upward of fifty years since its establishment, that which I first formed, called the Union Fire Company, still subsists, though the first members are all deceased but one, who is older by a year than I am. The fines that have been paid by members for absence at the monthly meetings have been applied to the purchase of fire-engines, ladders, fire-hooks, and other useful implements for each company; so that I question whether there is a city in the world better provided with the means of putting a stop to beginning conflagrations, and, in fact, since these institutions the city has never lost by fire more than one or two houses at a time, and the flames have often been extinguished before the house in which they began has been half consumed.

A major function of the fire departments in our cities is the inspection of buildings to enforce compliance with codes governing construction, maintenance, and occupancy. The first building codes, promulgated by municipalities in the late 19th century, were addressed to preventing conflagration, to keeping the fire from jumping to neighboring buildings. They specified materials for roofs and exteriors and the thickness and fire resistance of party walls. With the organization of the National Fire Protection Association (NFPA) in 1896, at the initiative of insurance companies, the building-code movement began to promote standard regulations covering a wider range of considerations. An NFPA publication in 1914 argued the case for limits on heights and areas for factory construction. That paper had to deal with special concerns raised by the longevity of 19th-century New England brick-walled textile-mill buildings with interior wood framing. Some of their wood internal members were of awesome dimension, and they had equally awesome heat content to contribute, albeit not too rapidly, to a hot fire. New factory building, by that time, had gone to masonry and steel construction with some claim to the label "fireproof," even with highly combustible contents. Sprinklers, first invented in 1852 and installed at an increasing rate in the classic New England textile-mill buildings late in the 19th century, now became mandatory in factory buildings. The codes gradually came to specify the performance of fire walls, separations between free-standing structures, protection at openings facing adjacent structures, provision of standpipes for hose connections inside buildings, storage of combustible materials, and so on. While fires inside buildings continue today to demand concern and action, the threat of urban conflagration under ordinary conditions of civil order has receded into the past.

An unanticipated, new threat is suburban conflagration in the states that share the Great American Desert. In California, especially, hillside real-estate

The San Francisco fire that followed the earthquake of 1906 is watched from one of the city's many hills. The conflagration escaped control by a fire department that had been crippled by the collapse of its firehouses during the first earthquake shocks.

developments leave the chaparral in place, to hold against erosion, right up to the backyards and house walls. A brush fire in Berkeley in 1923 destroyed 640 buildings. In 1961, a wind-driven conflagration rolled up the expensive Bel Air canyon development, destroying 512 homes.

Memories of the San Francisco earthquake in April, 1906, compel the further qualification that cities are not proof against the fire that must start on such occasions. April that year brought reminders all around the world of the forces at work in the crust of our planet. An earthquake in Formosa killed 1,000 people, a volcano erupted in the Canary Islands, Vesuvius had its biggest eruption since it destroyed Pompeii, and two substantial earthquakes were recorded in the Caucasus. In San Francisco, the stick-slip earthquake in the San Andreas Fault on the morning of April 18 overturned coal and wood stoves and kerosene lanterns and broke gas and water mains. Fires started simultaneously in wood buildings all over the city. With alarm lines down and firehouses collapsed (and the fire chief killed under one of these collapses), fires burning over the next two days destroyed 28,000 buildings across four square miles.

Geologists observe that cities holding an increasing percentage of the world's population stand on seismically active crust and that cities elsewhere are due to sustain earthquakes on only a less frequent schedule. In the words of Vladimir Keilis-Borok, at the U.S.S.R. Institute of Earth Physics, the vulnerability of these huge populations to the conflagrations that are bound to follow serious earthquakes under big modern cities portends catastrophe on the scale of the Black Death pandemic in 14th-century Europe.

A prevision of what might happen by fire in these cities is provided by the fire storms that attended the success of the Royal Air Force Bomber Command and the U.S. 8th and 20th Air Force in conducting incendiary raids on cities in Europe and Japan during World War II. The meteorological phenomenon called a fire storm has been observed on occasion in nature. At Peshtigo, Wisconsin, in October, 1871 (the same day as the Chicago fire), and near Sundance, Idaho, in September, 1967, forest fires occurred that developed into fire storms. Called a fire whirl when observed in uncertain development in smaller forest fires, it begins when the fire sends up a convection plume of hot gases that draws a flow of air radially inward at the bottom with sufficient force to dominate other local air circulation. Given the statistical instabilities in that circulation, the in-pouring wind pattern begins to develop a rotation, usually counterclockwise in the Northern Hemisphere. The result is a fire-induced cyclone, what one observer has called "a fire-generated rotating thunderstorm." It is similar in all other respects to the more common cyclones, tornados, and typhoons, and it generates the same enormous forces.

Forest fires each year burn through one to two million acres of the 225 million acres of U.S. forest land. Natural communities show evidence of adaptation to fire, with reduction of ground litter keeping fires out of treetops. The "crowning" of a fire, as in this photograph, spreads fire wildly.

The residents at Peshtigo reported a noise resembling that of a freight train building to a volume that made voice communication among fire fighters impossible. At Sundance surface winds were sustained at 50 miles per hour for nine hours, with peak winds estimated at 120 miles per hour. The fire moved 10 miles in that time and burned over 70 square miles of mountainous terrain. It is not clear that in that terrain the cyclone reached full development. There is no doubt about the fire storm at Peshtigo. It burned over 2,000 square miles of forest on relatively flat terrain, killing 800 people in the little town of Peshtigo and 1,500 more in the countryside around. This fire led to the creation of the National Forest System, with the prevention of such catastrophes among its missions.

At Hamburg, Germany, on February 27, 1943, after an incendiary-bomb raid by 1,000 bombers from both American and British air forces a fire storm burned out 3.2 square miles of the city and killed 21,000 people. It generated surface winds of 100 miles per hour to support its updraft. Casualties were found roasted in bomb shelters, and steel-frame buildings collapsed in the heat. At Dresden, from February 13 to 14, 1945, a similar prolonged raid set up a

A fire whirl develops in the deliberate burning of the litter remaining on a forest floor after clear-cutting of the timber. Many fires on the ground have here merged into a single column of fire. The convective updraft draws air into the fire horizontally around its periphery. Under the action of the Coriolis force, the column of smoke and flame has begun to rotate in the counterclockwise direction characteristic of Northern Hemisphere cyclones.

fire storm that burned out 4.6 square miles of the city and left 135,000 dead. Surface winds at 80 miles per hour were reported. In Tokyo on March 20, 1945, an incendiary attack by the largest number of bombers mustered for a single attack by the 20th Air Force set off fires that merged into a single conflagration that killed nearly 84,000 people. Surface winds of up to 70 miles per hour were recorded. Although the fire did not develop the full cyclonic pattern but rather seemed to roll forward on its course, it surely generated the hurricane wind that attended it.

At Hiroshima the fire storm developed about 20 minutes after the "prompt effects" of the first atomic bomb took effect. The bomb was of a size that made the radius of its blast and of its radiation effect coterminous from the altitude at which it was detonated, a radius of about one mile. The fire storm burned out a circular region of just that radius. The surface winds flowing radially inward peaked at about 40 miles per hour after about three hours and probably contained the fire within the radius of the prompt radiation.

Although most buildings in modern cities are fire resistant, their contents are not. Combustible materials are concentrated in them in much greater abundance than in cities before the Industrial Revolution. The fuel that has displaced human labor is present not only in tank farms and railroad marshal-

A Hamburg residential district in the aftermath of a fire storm ignited by a U.S.-British incendiary bomber raid in 1945 stands as utterly destroyed as Hiroshima after the fire storm ignited by the atomic bomb. The casualties in Hamburg were higher.

ing yards but in every factory, shop, office building, and household, in automobiles, and in neighborhood service-station tanks. Everywhere is the abundance of other material goods, almost all combustible, made or fashioned by mechanical energy. There is much more in modern cities to burn than before.

Calculations of these volumes of combustible material underlie the predictions that a "nuclear winter" might be triggered by the exchange of some 5,000 megatons of thermonuclear explosive power. Although the exchange ("strategic") is assumed to be directed at military targets on each side, the attacks must inevitably, if incidentally, find city targets. It is the dust and uncombusted soot lofted to high altitudes in the plumes of the fire storms in most large cities that would bring on the winter.

The effect of the soot particles, especially, has been calculated from large-scale fire experiments of the kind described in Chapter 5. The sheer size of the fires would create plumes high enough to pump hundreds of millions of tons of dust into the stratosphere and soot into the troposphere, where it would circulate around the globe, spreading to screen much of the Northern Hemisphere and perhaps some of the Southern. For the first six weeks as much as 99

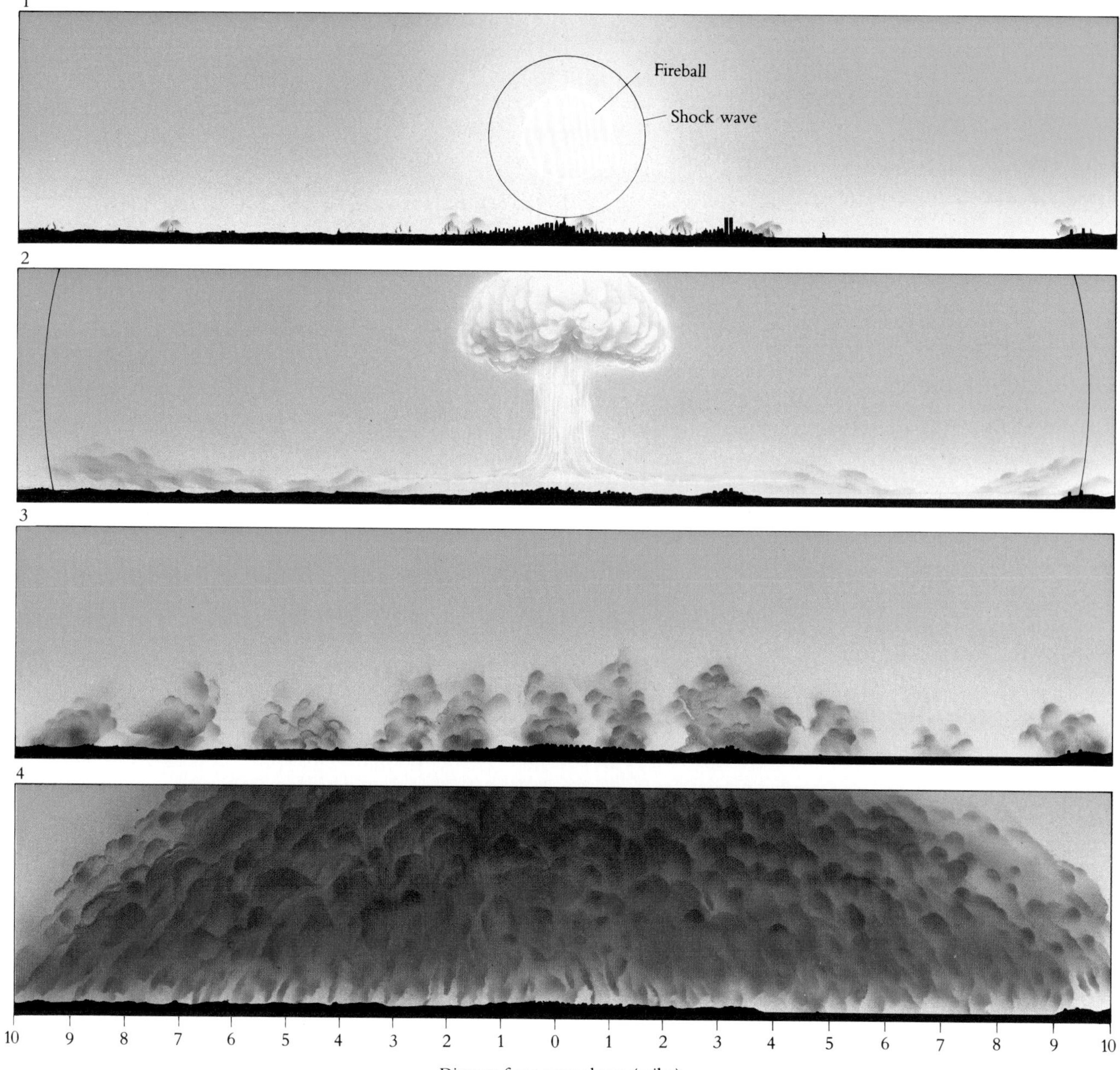

A fire storm develops in the aftermath of a one-megaton nuclear explosion over the heart of New York in the hypothetical sequence of events depicted. (The detonation point is assumed to be at a height of 6,500 feet directly over the Empire State Building.) In the first few seconds after the detonation the initial flash of thermal radiation from the fireball would spontaneously ignite fires in combustible materials at a considerable distance (1). Many of these fires would be promptly snuffed by the passage of the spherical blast wave (black arcs) and the accompanying high winds, but those two effects would also start a large number of secondary fires in the process of destroying most of the city's structures (2). Some of the individual primary and secondary fires could then merge into major conflagrations (3), which in turn might coalesce into a single massive fire covering most of the city. If such a fire were intense enough and the meteorological conditions were favorable, a full-scale fire storm could ensue, driven by winds of 100 miles per hour or more in the vicinity of the central updraft (4). Eventually the fire would burn itself out, leaving a smoldering residue. The smoke and dust thrown up by thousands of such explosions could extend over a large region, effectively blocking the sunlight and drastically reducing the surface temperature, regardless of the season.

The New York Crystal Palace was built in 1853 on the site "two miles from City Hall" now occupied by the New York Public Library and Bryant Park in midtown Manhattan. The largest building erected in the United States up to that time, it was to have housed a perpetual "exhibition of goods and articles from all parts of the world." A fire in an art exhibit brought it down in ruins in 1858. The vulnerability of "fireproof" iron construction taught the lesson that iron structural members must be insulated to keep their temperature below the 500°C weakening point of iron for the duration of any fire to which the building may be exposed.

percent of the solar radiation would be screened from the surface of the earth; after eight months, the calculations show only half of the solar radiation reaching the surface. Temperatures would drop accordingly by 20 to 40°C around the world and return to normal only after one year—at the lifting of the nuclear winter. This abnormal temperature excursion would have disastrous effects on the world's ecosystems, comparable to the aftermath of the asteroid impact to which the extinction of the dinosaurs has been attributed. Widespread concern has started up the development of global models that bring more parameters into calculation, with much of the data stemming again from research into unwanted fires.

The iron-framed, glass-walled and -roofed Crystal Palace exhibition hall erected in Manhattan in 1853 was hailed for, among other novel virtues, its imperviousness to fire. While under most circumstances glass and iron are noncombustible, the contents of the building were not. A fire starting in an art exhibition in 1858 brought the entire building down. The fire had raised the temperature in crucial structural members above the softening point of iron.

This dramatic lesson was not lost on builders, who were just beginning to rely on iron in building frames. They learned to insulate structural iron in casings of concrete, asbestos, or gypsum calculated to keep the metal below its 500°C softening point for the duration of a fire. Yet the same lesson had to be taught again as late as 1967, with the erection and collapse by fire, within a

The wilted steel framework of the McCormick Place exhibition hall, destroyed by fire in 1967, testifies that the lesson of the fire that destroyed the Crystal Palace in New York City in 1858 was not learned. Building codes require the insulation of a steel structure to keep the temperature below 500°C, the softening point of steel, for the duration of any fire to which the building may be exposed.

year, of the McCormick Place exhibition hall in Chicago.

Frequent fires in the tenements that housed the impoverished immigrant population in New York City began the teaching of another lesson not yet adequately incorporated in some buildings under construction today. This is the need to provide the inhabitants with a secure escape route. A tenement fire in 1860 in which 20 people died prompted an ordinance requiring that all new housing for eight families or more be made fireproof or be equipped with iron fire escapes attached to the outside walls. When these adornments pulled away and fell from the walls in fires, builders of sufficiently large apartment houses were compelled to build them of masonry and steel and to provide fully enclosed stairwells separated from the rest of the building by fire doors.

Those New York City ordinances did not cover commercial establishments. In 1911 the scandalous Triangle Shirtwaist fire took 146 lives, principally of young immigrant women. The building was not fire resistant in structure, had no sprinklers, and provided no fire escapes.

In 1912 the fire that destroyed the enormous 10-story office building of the Equitable Life Insurance Company painfully taught and retaught some more lessons. Comprising five different buildings built at different times and joined together, principally by facade, this building rested in part on unprotected cast-iron columns. Neither stairwells nor elevator shafts could be sealed off. The rooftop water tanks held a water supply inadequate to suppress a fire, and the nearest high-pressure water supply on the ground was 800 feet away. With no strategy for coping with a 10-story fire, the city's fire department could not help. The building was destroyed; chimney effects in the stairwells and elevator shafts spread the fire throughout, and the cast-iron columns collapsed. Fortunately only six lives were lost.

In 1926 another fire occurred in the rebuilt Equitable building, this time on the 35th floor. The fire was severe, but the structure remained intact. Water from the building's own supply put the fire out. No one was killed.

The building codes of the United States, under the jurisdiction of local governments, reflect a uniformity promoted by the insurance industry and by the inducements and sanctions available to the Federal government. Fire remains their principal concern (structural integrity and sanitation being the others), and measures to prevent fire or minimize its consequences constitute most of the text. These codes proceed from the reasonably secure premise that public buildings in this country are now of fire-resistant construction. They provide for the safety of occupants in a fire mainly by the careful specification of interior passageways, stairwells, and distance to and number of doors. To keep the stairwells from becoming chimneys and passageways for fires and ensure that people can use them for escape, the codes say how many stairwells there should be per square foot of floor space and where they are to be placed, prescribe the performance of their doors to permit access by people and keep fire and smoke out, and specify the materials of construction.

Problems arise and catastrophes happen when the codes are not followed. Hotel fires have produced heavy casualty lists. In the 47-year period from 1934 to 1961, 130 hotel fires killed 1,204 people. In the MGM Grand Hotel fire at Las Vegas, Nevada, in November, 1980, 84 people died. The investigator's report showed that in this hotel, built in 1972–73, three of the four stairwells "were not enclosed in two-hour fire-rated construction . . . [and]

Fire escapes on New York City buildings were made mandatory in the mid-19th century after fires in tenements scandalized the community with their death tolls. Better than no alternative escape route, they are subject to loss of strength in the heat of a fire and tend to pull away and fall from the building. Fire-protected inside stairwells have replaced them.

Major U.S. Hotel Fires

Hotel	Date	Lives lost
Winecoff, Atlanta, Ga.	1946	119–122
LaSalle, Chicago	1946	61
MGM Grand, Las Vegas, Nev.	1980	84
Hilton, Las Vegas, Nev.	1980	8
Stouffers Inn, Harrison, N.Y.	1981	26

The MGM Grand Hotel fire in Las Vegas, Nevada, in 1980 killed 84 people, largely by the spread of combustion gases up stairways and vertical shafts. As in all such disasters, hindsight showed the construction to have been below the standards of the codes that keep fires confined in properly designed and constructed buildings.

nonrated access panels allowed fire and products of combustion to spread into these stairways."

A special problem for code writers and enforcement officials is presented by theaters and dance and music halls that crowd large numbers of people into enclosed spaces for entertainment. The structures themselves have at times proved not to be fire resistant. The aftermath of fires has shown inadequate provision and malfunction of exits. What makes for large casualty lists is the high surface-to-volume ratio of the mass of material, of whatever kind, that goes into the sets of theaters and into the decoration and furnishing of the dance and music halls.

The archetype of these fires (*see the table on the opposite page*) is that in the Cocoanut Grove nightclub in Boston in November, 1942. The fire began in the basement lounge and was carried by the combustible decor up the stairs into the main chamber. Behind the principal exit, a revolving door, 200 dead were counted. Another exit opened inward; there 100 dead were counted. Other exits were locked. Altogether 492 died, nearly half the crowd packed into the place that evening. Many more were severely burned. The episode provided an unexpected exercise for the civil defense organization and procedures set up in Boston as in other East Coast cities in the early months of World War II. More blood plasma was used in treating the casualties than had been at Pearl Harbor, the year before.

Major U.S. Fires in Places of Public Entertainment

Location	Date	Lives lost
Iroquois Theater, Chicago	1903	602
Dance hall, Natchez, Miss.	1940	207
Cocoanut Grove nightclub, Boston	1942	492
Ringling Bros. Circus, Hartford, Conn.	1944	163
Beverly Hills Supper Club, Southgate, Ky.	1977	164

Against such eventualities the codes now specify the number and performance of exit doors. Signs posted in all such places, large and small, declare the number of persons in excess of which occupancy is dangerous and unlawful.

The fires that break out in these crowded establishments overtake people who, for the most part, are in possession of full adult physical and mental capacities. In distressing contrast, fires in hospitals and nursing homes trap people whose disabilities make them wholly dependent on attendants and fire fighters for survival. Nursing-home fires make the news in shameful recurrence every winter (*see the table on page 126*) because these enterprises so often find their housing in old buildings rarely intended for such a purpose and often not updated to comply with building codes. Even when they occupy reasonably modern buildings, as in the Marietta, Ohio, case, tragedy betrays a lack of sprinklers and automatic fire-door closers. The Marietta fire in 1970 was spread by combustibles in corridors. Most of the victims were asphyxiated in their beds, not burned.

Fires of this kind teach several lessons that are implicit in the strategy that looks to the containment of fire when it gets started in a modern building and to the protection of the occupants there. The first is that evacuation of people from a burning building is not always possible. This may be the situation not only of bedridden patients in a one-story nursing home but of office workers in a skyscraper and of guests in a multistory hotel. The fire-protection system must be predicated on the strategy of defending the patients, office workers, and hotel guests in place.

Second, it is not enough to keep the fire itself from spreading; its combustion products can be lethal at a distance. The gases penetrate every nook and cranny and can enter a room under a closed door and reach its victims. Heavy, sealed windows block access to fresh air.

The third lesson concerns people's behavior. Those responsible for the safety of others must understand the dynamics of the fire and the spread of its

Fire Disasters in Health-Care Facilities

Date	Building	Location	Fatalities	Notes
1918	Grey Nunnery	Montreal	53 (babies)	Wood interior
1918	Oklahoma State Hospital for the Insane	Norman, Okla.	38	Combustible interior
1923	Illinois State Hospital for the Insane	Dunning, Ill.	18	All wood buildings; doors opened inward
1927	Hospice St. Charles	Quebec	37	Wood interior
1949	St. Anthony's Hospital	Effingham, Ill.	74	Combustible interior; open stairways; no fire barriers
1957	Nursing home	Warrenton, Mo.	72	
1961	Hartford hospital	Hartford, Conn.	16	Combustible ceiling, tiles, and wall coverings
1963	Nursing home	Fitchville, Ohio	63	
1970	Nursing home	Marietta, Ohio	31	Combustible corridor linings
1980	Hotel (largely mentally retarded and mentally ill outpatients)	Bradley Beach, N.J.	24	Open stairs; single exit from upper floors

lethality. Building occupants must learn that evacuation is not the only way to safety and may not offer the surest way.

This strategy rests on understanding the way a fire develops and moves in a large building. The vast store of information from the field accumulated in the experience of the fire services laid the foundation for such understanding. Test fires in the model room and in the room-corridor and even the room-corridor-room configurations have made their contribution, as recounted in Chapter 5. A much smaller amount of data has come from test fires in large buildings conducted when an opportunity was presented by their scheduled demolition. Careful and extensive studies have been made of the movement of gases in functioning buildings, using nontoxic gases or harmless smoke bombs. On rare occasions, an actual fire has been set in a protected special area to study gas movement.

Chopping a hole in the roof is a fire-fighting tactic confirmed by experimental study of the spread of fire through a building from its room of origin. The hole can vent smoke laden with uncombusted fuel gases and particles that would burn explosively if it stayed hot and encountered air inside the building.

The lore and the understanding go to endorse the tactic of the fire fighter who chops a hole in the roof to vent the fire to the outside. If all goes well, the gas flow through the vent will take the uncombusted fuel gases outside and keep them from finding their way into other parts of the building. Inside the building more cold air flows through the corridor, less gas accumulates under the corridor ceiling, less heat radiates to the floor, and cold air reaching the fire room may slow the growth of the fire. The atmosphere inside the building clears as smoke exits through the roof, and the fire fighters can see better.

The only danger in this tactic is that the vent might cause an explosive fuel-oxygen mix to form in the fire room. This might occur if fluctuations in the chaos of the fire cause more air to enter the fire room. The force of the explosion from such a backdraft may be enough to knock down the building.

Tibor Harmathy of the National Research Council of Canada has recommended that architects incorporate this tactic, which he terms the "fire drain" approach, into the structure of the building. Windows, he says, should be made to open all the way to the top, leaving no trap under the ceiling to accumulate the gases. Recent work at the Factory Mutual Research Corporation and in the Division of Applied Sciences at Harvard University has shown that removing the wall over a doorway only modestly dissipates the smoke and gas layer. In one instance, the effect was to intensify the fire with an increased supply of air. It must be that a doorway or window wider than usual is needed to draw off the hot layer; experiments have not yet been run to fix the critical width.

The movement of fire and smoke in a tall building is much affected by differences in the density of the air inside and outside the building, created by differences in temperature, especially at their winter and summer peak. In winter, if the building is vented at the top, the air will rise rapidly, as in a chimney. In summer, with air conditioning, the opposite might occur. Such drafts may be substantial. For a 60-meter or 20-story building at a temperature difference of 40°C (+20 degrees inside and −20 degrees outside), there will be a pressure difference at the bottom between the outside air and the air inside of 55 pascals (0.2 inch of water pressure), the pressure outside at the bottom being greater than that inside, and of −55 pascals at the top floor (with a plane of neutral pressure about halfway up the building). At such a difference sudden venting at the top would bring a considerable rush of air in at the bottom, given an opening there, too. A comparable draft would be fine in a chimney, but it can carry surprise air flooding through an entire building if it has not been anticipated in the design of the air-handling system.

The inside-outside pressure difference—the stack effect—must be reckoned with in the strategy of smoke control. In a number of buildings smoke-control systems have been installed to pressurize stairwells and, in a few buildings, to make it possible to pressurize some parts of the building while depressurizing others. In the United States the systems tend to employ the heating, ventilation, and air-conditioning ductwork; in Japan the systems have separate ductwork.

The pressurizing of stairwells is meant to provide a smoke-free escape route for occupants. There must be a fire door at each floor of reasonably tight fit and a pressurizing fan capable of exceeding the stack effect plus the additional pressure exerted by the fire at the fire floor. A fully developed room fire may develop pressures of 25 pascals (0.1 inch of water pressure), which may in some cases be fully additive to the stack pressure depending on the floor of the fire and whether or not the air outside the building is colder or warmer than that inside. One may need to develop as much as 50 to 60 pascals of pressure inside the stairwell to hold back the hot fire gases. This has been tried, and it does indeed work. The doors become difficult to open, however, if the pressure rises too high. At 50 pascals (0.2 inch of water), about 11.4 newtons (25 pounds) of force would be needed to push open the fire door.

Building owners in New York City protested an ordinance that required stairwell pressurization, and so the ordinance was not enforced.

A more sophisticated scheme is to divide a building into zones horizontally and provide each zone with both pressurization and exhaust options. If a fire occurs on the ninth floor (*see the figure on page 129*), that zone is put on exhaust to the outside, tending to make its pressure negative with respect to

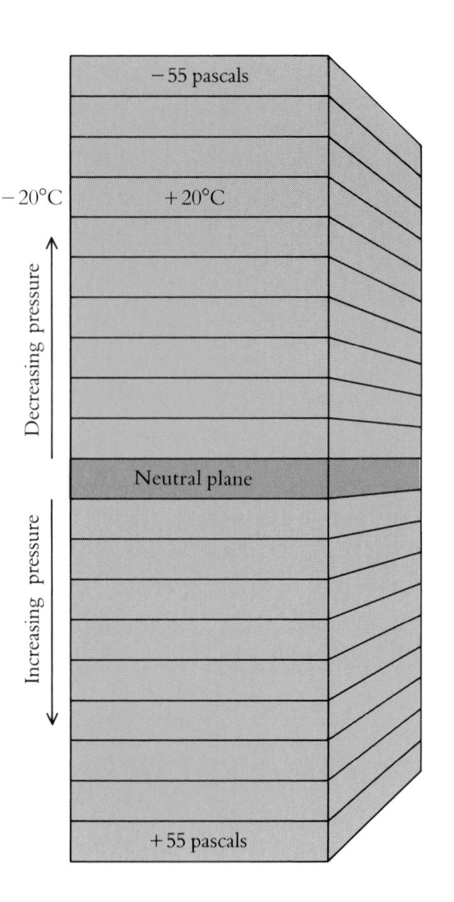

The stack effect in a tall building can set up a strong vertical draft if it is not properly managed in the design. Such a draft could spread fire from lower to upper floors. The effect is caused by the temperature differences inside and outside a building and the normal change of pressure with altitude: for a 20-story building and a 40°C temperature difference (with the inside warmer) this is a positive outside pressure of 55 pascals at the ground and a negative outside pressure of 55 pascals at the top, a 110-pascal gradient, compared with the much gentler slope of a barometric gradient. At about half of the building's height, the inside and outside pressures are neutral.

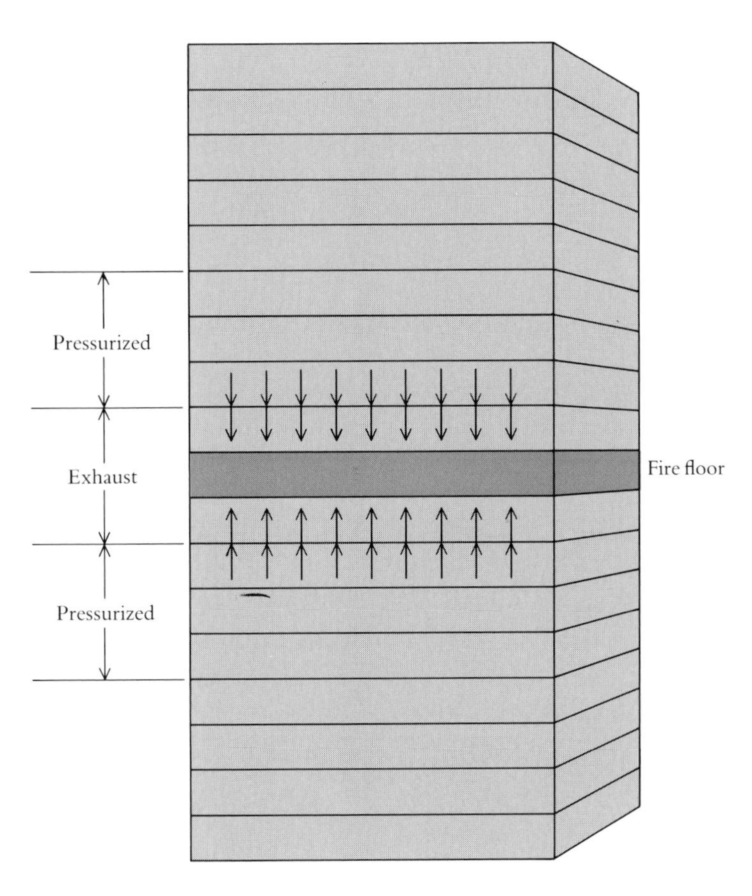

One strategy for fire fighting in a tall building employs positive and negative air pressure to overcome the stack effect (see the figure on page 128) and confine the fire to its floor of origin. The building air-conditioning system would be used to exhaust air from the fire floor, keeping the buoyant combustion gases from traveling inside the building. Positive pressure on other floors, especially on the floors above the fire, would keep those gases out.

adjacent zones. The neighboring zones are pressurized slightly in order to increase the pressure difference with respect to the fire zone. Smoke and gases are pumped to the outside and the occupants are advised to move to the adjacent zones. Ideally, all this may someday be handled routinely by computer controllers.

The beginnings of such systems have been installed in a few buildings, and the concept has been demonstrated successfully. Of course, the installations cost money. In the Seattle Federal Building, however, it has been claimed that trade-offs with other fire-safe design features brought the installation in at negligible increase in overall cost.

However absolute the value attached to human life, such a fire-safety system is a cost to be balanced with other economic considerations. Building codes are a patchwork of requirements, each added after some large fire disas-

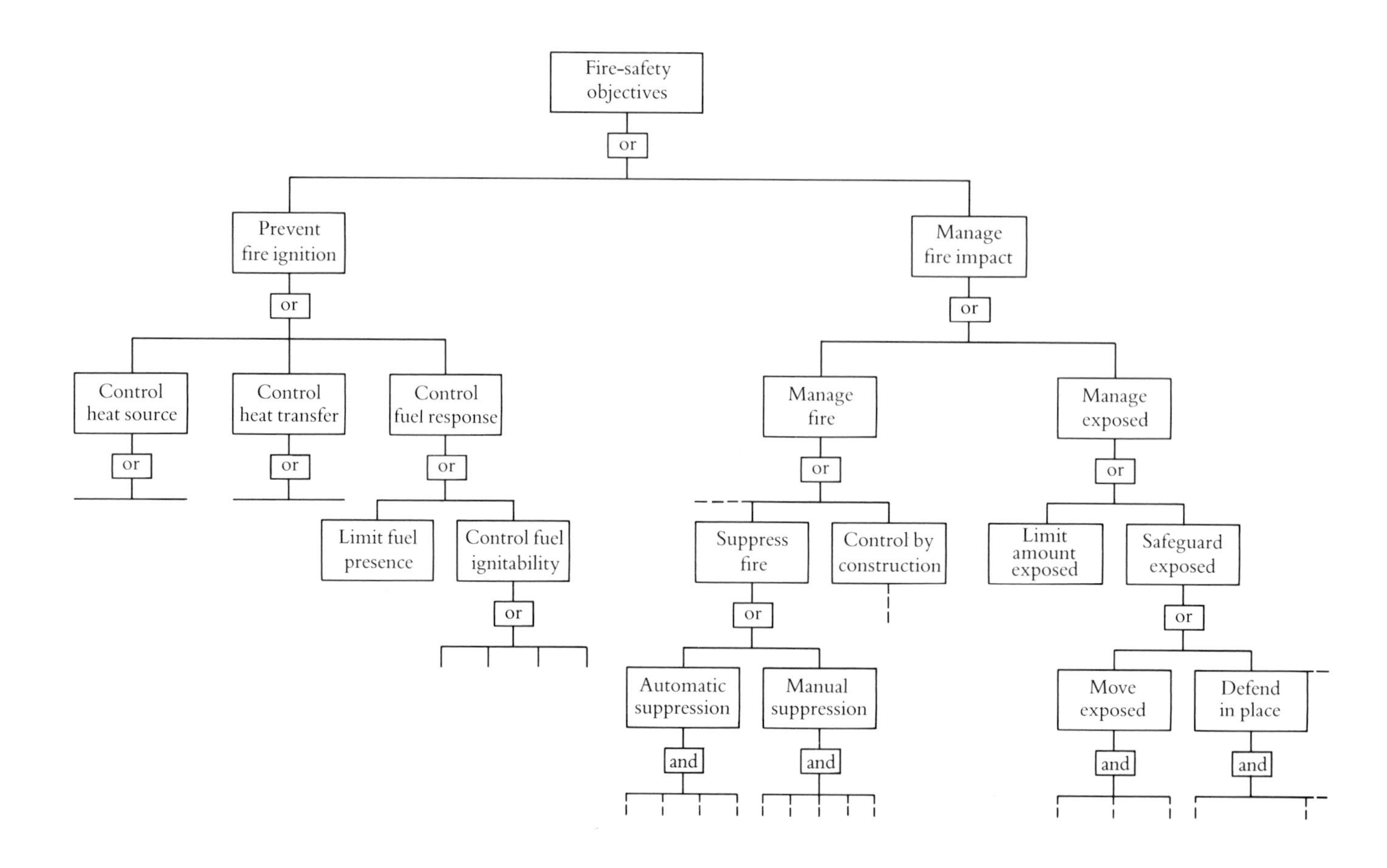

ter. In many cases they add unnecessarily to the cost of a building. Some provisions are redundant; the requirement of both sprinklers and fire-rated walls in an office area, for example. Recognizing that they cannot remove all chance of fire, safety engineers attempt to reduce the chance of the growth of a fire beyond a certain size to a suitably low probability.

The quantification of risk is an exercise not appreciated by the public, even though people take chances with their lives every day. Architects and engineers must nonetheless make rational judgments, calculations, and choices. They cannot design for each of the endless combinations of events that can produce a serious fire. They must design for the most likely scenarios and put safety resources where they will do the most good. Considerable care and cost are lavished on preventing electrical short circuits by making secure electrical connections and insulating the surroundings from overheating by stoves, heaters, fireplaces, and furnaces. On the other hand, an occupancy

A decision tree developed by a committee of fire technologists organizes the approach to preventing and to fighting fire by both passive and active measures. With the tree in mind at the time of construction of a building, important steps can be taken under "manage fire" to "control by construction." If, on the far right, "move exposed" is to be an option, that too must have been anticipated in the design of the building. On the other hand, the tree prompts active measures to prevent ignition; for example, "limit fuel presence." In its entirety the size of the above tree runs on almost three feet from the first branches shown here.

designed for work may not require the same alarms as those necessary in occupancies designed for sleeping.

For the past 15 years engineers burdened with the responsibility for trade-offs between increments of safety and increments of cost have put their minds to development of a systematic approach to sorting out the alternatives. The public had become alarmed, especially after the Marietta nursing-home fire, at the relatively low level of fire safety found in such facilities. Owners protested that the cost of raising the level of fire safety would be prohibitive if the prescriptions of the pertinent codes were followed. The codes do allow alternative approaches, however, if they provide equivalent safety. The problem was that there was no agreed-on way to determine this equivalence. Fire-prevention engineers first developed a method to represent the logic of fire-safety measures prescribed by a code or by a design. The result is a decision tree that enables one to see at a glance all the factors that must be considered in setting up a design strategy. This analysis has been adopted by a national fire protection association committee and is revised periodically. The complete version, printed in readable type, makes a scroll nearly three feet long by 11 inches high. A condensed version is shown in the figure on the opposite page. In the figure there is a series of branches, some preceded by the symbol for an "or" gate, meaning that one proceeds along one branch only, and some preceded by the symbol for an "and" gate, meaning that both or all branches must be followed and the prescriptions observed. At the top of the tree, ignition is either assumed to occur and the impact of the fire must be dealt with by the designer, or not to occur because the measures prescribed on this branch of the fork have all been faithfully taken.

Sustained ignition can be prevented, according to the diagram, by controlling the heat source, by interfering with heat transfer, or by controlling the fuel itself. Each of these alternatives leads to subalternatives. The fuel can be controlled, as shown in the diagram, by limiting the amount of combustibles in the space, by specifying a limit on the ease of ignition of the fuels such that the most probable heat sources will not ignite the allowed fuels, and by four other alternative measures listed under this choice.

If it is assumed ignition has occurred, the right-hand side of the decision tree provides two alternatives for responding: manage the fire itself or manage the people exposed to the fire. One can manage the fire by controlling the type and amount of fuel present, by the design of the structure itself, or by suppression. Control of fuel applies to such items as interior finish (walls and ceiling coverings), carpeting, and furnishings. (None of these are controlled in private dwellings.) Control of the structure refers to passive items such as fire-rated walls, ceilings, floors, doors, columns and beams, and provisions of fire stops in walls and smoke breaks in corridors and of enclosed stairways, and so on. These things are done in the design phase and are essentially unalterable features of the building. There are also some active construction features that control fire movement. These include automatic door closers, smoke-control and exhaust systems, pressurization systems, and fire vents.

The suppression option includes automatic systems (usually water sprinklers) and manual ones (usually the local fire department). Either technique requires a means of detecting the fire and a means of activating the suppression apparatus. For the automatic sprinkler the detector is also the activator. A link that melts at low temperature (thereby detecting the fire) opens the water valve and suppression begins. To summon fire fighters one must both detect the fire and send the alarm signal to the fire station.

Some of the techniques used for managing a fire are also essential for managing the exposed occupants. Fire and smoke detectors sound the alarm for occupants. In addition certain communication devices can be provided to tell occupants where the fire is, where they are to go, and what they are to do. Passive construction features provide the necessary escape routes or areas of refuge. Smoke-control and pressurization systems keep the fire away from these areas and provide a tenable environment for refugees.

Designers and code officials use the tree to help with their review of plans to make sure all possibilities have been considered. The tree does not tell them what level of fire safety is provided by a given design because the tree is not quantitative. Researchers had to come up with something more to make it possible to compare designs with one another or with provisions of a code. Stimulated by Congressional concerns about nursing-home fires, the then Department of Health, Education, and Welfare asked a team at the National Bureau of Standards to define the level of safety required in the codes and to show the equivalence of alternative designs. As a model they adopted the National Fire Protection Association Life Safety Code, which had been made mandatory for all nursing homes receiving any form of Federal assistance. The code was analyzed, and 13 safety parameters were found to account satisfactorily for the required safety level. These could be grouped together in different ways to address three concerns: fire containment, fire "extinguishment," and the movement of people.

The 13 safety parameters in the Bureau of Standards system are shown in the figure on the opposite page, some applying to more than one concern. It can be seen that automatic sprinklers affect all three of the safety concerns listed above, smoke detection and alarm systems address two of the concerns (extinguishment and people movement), and the nature of the basic construction addresses two of the three. The blocks in the figure are filled in for each zone of the building (usually a floor or section of a floor) using numbers set by a committee of fire experts for each specific condition. Thus, for construction, the value to be entered in the two blocks varies from −13 for unprotected wood-frame construction at the fourth floor or above to +4 for fire-resistive construction, designed to withstand a specified fire exposure. Materials are still rated on the basis of their performance in the test of exposure to a fire in which

Safety parameters	Containment safety (S_1)	Extinguishment safety (S_2)	People-movement safety (S_3)	General safety (S_G)
1. Construction				
2. Interior finish (corridor and exit)				
3. Interior finish (rooms)				
4. Corridor partitions or walls				
5. Doors to corridor				
6. Zone dimensions				
7. Vertical openings				
8. Hazardous areas				
9. Smoke control				
10. Emergency movement routes				
11. Manual fire alarm				
12. Smoke detection and alarm				
13. Automatic sprinklers				
Total value	$S_1 =$	$S_2 =$	$S_3 =$	$S_G =$

Quantification of the evaluation of safety is the object of this checklist. The measures listed in the column at the left bear variously on the objective of containment or extinguishment of the fire and on the movement of people to safety. Thus measure 5, "doors to corridor"—their number, placement, and design—is essential to containment of the fire and the movement of people to safety but has little to do with extinguishing a fire. Numerical values assigned to each measure yield a total within which trade-offs, as, for example, between sprinklers and construction measures, may be managed to secure the same safety objective at somewhat lower cost.

they serve as the ceiling of a 25-foot tunnel. In the tunnel test, discussed in Chapter 5, asbestos board receives a 0 flame-spread index and red-oak panels receive an index of 100. Values for smoke detection and alarm vary from 0 for none through 4 for installations in all corridors and habitable spaces to 5 for detectors in all spaces. The highest value is for automatic sprinklers; sprinklers in all spaces receives a +10.

As is apparent, the assignment of numbers in each category reflects an estimate of the contribution of the category to the containment and suppression of fire and to survival by occupants. For nursing homes the risk in each category must meet a minimum numerical standard; the sum of the three brings the ensemble well above the minimum grand total required for other occupancies.

At present, for other purposes, the three columns are summed individually, the rows are totaled, and the last column is then summed. The four totals are compared with the levels calculated from the code. If each code level is exceeded, the building or building design is deemed acceptable. If one of the first three is deficient, one of the parameters contributing to that sum must be adjusted upward. In the prescriptive approach, the designer simply uses the prescribed design of the code. In the Bureau of Standards scheme, the designer may adjust any one of the parameters within the family contributing to that subtotal. Clearly more flexibility is permitted the designer and the public official who must approve the design. There is opportunity to select options that affect both economics and aesthetics.

This approach to fire-safety design is at present limited to health-care facilities and other special occupancies, since the current version was tailored to the type of interior furnishings and probable fire scenarios encountered in such buildings. The National Fire Protection Association has incorporated the method as an appendix to the Life Safety Code and the Department of Health, Education, and Welfare (now the Department of Health and Human Services) currently accepts this as an alternative design approach for facilities in its domain. The system is being adapted to other occupancies in the immediate future. It is flexible and readily expandable. It is now being put in place in care facilities for the physically and mentally handicapped, in correctional facilities, and in buildings in the national parks. Details regarding furnishings in hotels and motels, theaters, airports, and the like can be added.

It must be noted that the numbers intended to quantify value and risk in the present procedure reflect intuitive evaluations. It is hoped that the experience that goes into these intuitive judgments will be fortified by the results of large- and small-scale fire tests and the mathematical models that are being developed from them. When the fire models are completed and validated, we shall be able to provide some absolute numbers for at least parts of the comparison, those parts dealing with fire containment, for example. It is likely that some parts of the evaluation will remain intuitive for several more years. The quantitative correlation of such factors as length and width of corridors with risk to life is a problem on which little progress has been made. Indeed, well-controlled studies of the movement of people under stress of extreme danger are impossible to conduct.

Building codes, even when they are perfected, cannot always address the fire worthiness of products brought into the structure by occupants. For example, the furnishings brought by apartment dwellers cannot be controlled by the owner of the apartment or the municipality. If I want to fill my apartment with highly combustible items, that is my affair. Protecting occupants from fires starting in their own premises is therefore not a job for the code for large buildings. Rather, the code is designed to protect my neighbors against my recklessness.

Beyond the perfection of the computer models comes the question of how to incorporate them in the building regulatory process. The issue is broader by far than just how to reference a computer model. What is really happening is that the computer is challenging the logic and internal consistency of the codes. Perhaps sometime in the next 20 years or so our codes will be completely redone using the tools of computer science. To do this requires, first of all, that the components of the codes be logically written and rigorously tied together. The properties of materials will have to be measured scientifically (quantitatively and reproducibly). Test methods must be thoroughly understood so that exceptions to the results are predictable and can be automatically set aside. Models of fire and smoke growth and spread must be able to operate entirely on test results, geometry, and occupancy information to produce estimates of escape times. Standards invoked by codes will be written in machine-readable logic. The code itself will be computerized.

Ideally, then, it should be possible for a builder to ask a code official if he may use material X on the walls of corridors in a high-rise apartment building. The code official elicits from the builder a set of design data (how much of X and where) and properties for material X. He puts them into his computer along with key facts about occupancy and overall building design. The computer then tells the code official probable escape times for the most likely fire scenarios and indicates whether the critical hazard is heat, smoke particles, or toxic gases.

There are two variations of this approach. One is fairly inflexible: the builder consults the official only at the start and end. Alternatively, the builder interacts with the official and makes use of his expert knowledge, added to data put into the system from one or more other experts. Such an "expert system" installed in computers is now recognized in medical diagnosis, in troubleshooting in diesel engines for locomotives, and in designing parts of complex computer systems. It is my view that the computerized expert-system approach to building design and regulation is the next major development in building codes. To bring it into play, the full computer model must be perfected, the necessary data supplied from fire research, and the experience of experts mixed in thoroughly. Then, of course, it will be necessary to raise the consciousness and computer awareness of the entire building community. This is not an entirely romantic vision. Architects, engineers, and code officials already use computers to check the structural design of buildings.

An explosion of single-family dwellings near Phoenix, Arizona, is shown. This mode of housing gives Americans the world's largest per capita energy expense and increases their exposure to the risk of fire. It is the two-thirds of all fire deaths incurred at home, half of these from fires started by cigarettes, that gives U.S. and Canadian citizens the worst fire records in the world. Urban slum dwellers, however, are still the major victims of fires.

7 *House Afire*

With the measures taken thus far to suppress the menace of fire, it may be said, society has succeeded in protecting people from the misfortune or carelessness of other people. Conflagration is now only a lurid memory, except as incidental to such catastrophes as earthquakes, volcanos, and war. Skyscrapers, hospitals, and apartment houses are today largely fire resistant. In these buildings, a fire can be contained in the room or floor of its origin and the circulation of its toxic smoke localized. When building regulations have been followed, people are reasonably secure from fire away from home.

Further major inroads on the loss of property and life to fire must come from suppressing the menace in the home. As the figures cited at the outset of this book show, 80 percent of the people killed by fires are at home or at leisure elsewhere in hotels or other public places; a full 80 percent of these are killed in their own homes. The complicity of the victims' behavior in these numbers is indicated by some more statistics. To begin with, 27 percent of all fire deaths start with a cigarette.

In higher percentages than in other urbanized industrial countries, people in the United States and Canada live in separate, free-standing residences. In their continuing migration to suburbia and beyond, they are living in their own homes in increasing numbers. They seem thereby to have increased their risk of exposure to fire. Suppression of the menace of fire to these people has become, therefore, the delicate task of protecting them from themselves.

Beginning in the late 1960's, U.S. citizens gave signs that they were disposed to undertake that delicate task. Through their representatives in Congress, they brought fire forward as a public issue. Hearings were held, visits to hospital burn wards made, and studies commissioned. As the yield from this activity, Congress passed four laws in the seven years from 1967 to 1974 dealing with fire prevention and control.

The first piece of legislation recognized the role of textiles as the first item ignited, for example, clothing, by stoves, space heaters, or matches, or as the means of the spreading of fire, for example, across carpeting, up drapes, over upholstery, or via bedding. This 1967 amendment to the Flammable

*The Top Fire Death Scenarios**

Rank	Occupancy	Item ignited	Ignition source	Percent of U.S. fire deaths
1	Residential	Furnishings	Smoking	27
2	Residential	Furnishings	Open flame	5
3	Transportation	Flammable fluids	Several	4
	Independent (residential)	Apparel	Heating and cooking equipment	4
	Residential	Furnishings	Heating and cooking equipment	4
6	Independent	Apparel/flammable liquids	Several	3
	Residential	Flammable liquids	Heating and cooking equipment	3
	Residential	Flammable liquids	Open flame	3
	Independent	Apparel	Open flame	3
10	Residential	Interior finish	Heating and cooking equipment	2
	Residential	Interior finish	Electrical equipment	2
	Independent	Apparel	Smoking	2
	Residential	Structural	Electrical equipment	2
	Residential	Trash	Smoking	2
				66
Others, all less than 2 percent of total				34
				100

*From Clarke and Ottoson.

Fabrics Act of 1954 extended legislation that had been passed in the Eisenhower administration to protect children from certain egregiously flammable materials employed in such novelty items as cowboy costumes. Going beyond consumerist agitation, the 1967 legislation also sought to enlist manufacturers in protecting people from themselves. Procedures for establishing Federal standards of flammability were provided, and the administration of them was later vested in the Consumer Product Safety Commission, created by legislation in 1972.

The Fire Research and Safety Act of 1968 empowered a national commission to conduct a comprehensive study of fire and initiated technical enterprises in Federal agencies, including some of the research reported in these pages. From the work of that commission came the Federal Fire Control and Prevention Act of 1974. Here Congress charged the Federal executive agencies to provide the data, technology, and other tools to support the work of local fire-protection agencies. This legislation, adopted at the crest of the they-ought-to-pass-a-law sentiment about fire safety, firmly declares that the burden of protecting people from fire rests with local governments, their building codes and fire departments.

The thrust of building codes with respect to the unattached family home is to control the fire hazard in those elements in the structure that are hidden from the buyer and occupant, that require considerable professional expertise to assess, and that are difficult to correct once the construction is completed. Examples of such regulation are the requirement that fire stops be placed between vertical studs in the wall of a house (no longer of as much moment, since outer wall spaces have come to be filled with insulation), the design of fireplaces and chimneys to keep them from overheating the combustibles in the surrounding structure, and control of electrical work to prevent short circuits in the walls. These minimal provisions are enforced by periodic inspection during construction, for mistakes not caught and corrected could then be impossible to detect afterward. To trigger this proceeding most jurisdictions require building permits even of people building or making major changes to their own house on their own property.

The codes for one- and two-family dwellings do not address such important matters as openings between floors, the fire rating of flooring and carpeting and of wall and ceiling finish, or the placement of fire-rated barriers around space heaters. All of these and still other aspects of design are specified in the codes that govern buildings open to public access and occupancy. Nor do the family-dwelling codes require the presence of fire-fighting equipment, standpipes and hoses, extinguishers or sprinklers. Until recently, those codes made no mention of fire-detecting equipment.

Without arguing that fire or smoke detectors in the home are necessary for the good of neighbors, some local legislators have been imposing this measure of self-protection on homeowners. In most jurisdictions such installation is required at the time of the building or the impending sale of the house. In some, however, homeowners are required to install these detectors simply for their own protection.

It is a curious fact that public regulation, some of it stemming from Federal action, has now made the so-called mobile home safer from fire than the full-size family dwelling. Specifications—of furnishing materials, of the insulation of the cooking range from the structure, of door openings and access, and so on—no one has dared to suggest to the building industry or the

The contrasting flammability of untreated textiles (this page) and treated textiles (opposite page) used in children's clothing is demonstrated here. In both cases the ignition source is a small flame, accepted in the Federal standard for such tests.

homeowner have been imposed on the fabricators of mobile homes. Since these now constitute 25 percent of the new housing in this country, that means many families are safer than they might otherwise be.

Fighting a fire in an unattached dwelling does not fit the tactics and strategy accepted for the control of fire in a large building. The fire cannot be permitted to proceed to burn out of the room in which it starts. There is no way to contain it after that. In bursting out at flashover into the rest of the floor, the fire would very likely kill or seriously injure every occupant of that floor and probably on the floor above. The strategy for large structures is plainly unsuitable for small dwellings. Indeed, many still question the adequacy of that strategy except as a protection for other people elsewhere in the building.

Nonetheless, the research and thinking that have gone into the control of fire in big buildings have their relevance to the control of fire in the unattached dwelling. In particular, the decision tree developed to organize the approach to fire control in those buildings can help to organize one's thinking about fire in one's home. From that one-foot by three-foot scroll (*see the figure on page 130*), we can extract the salient points and choose the measures it would be sensible to take in any household.

At the first branching point in the decision tree is "ignition" or "no ignition." Election of the second alternative suggests a checklist of common-

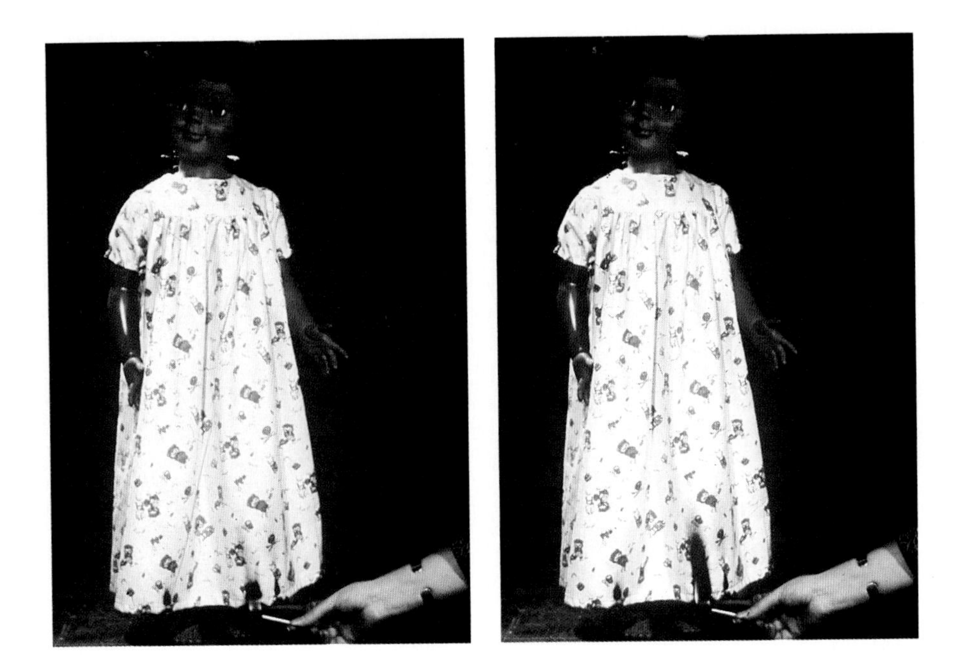

sense measures. One does not store gasoline inside the house; the propane tank is installed outside. One does not run a stove pipe through a wood wall or hang curtains over a gas stove. Prevention of ignition also compels consideration of the site of first ignition. This, as we have noted, is often some textile or other, especially for fires in the home. The flammability of textiles is a topic that takes us into interesting applications of research into the physics and chemistry of the flame and into some interesting aspects of U.S. politics.

There are now five Federal standards governing the ease of ignition and the burning rate of textiles established under the Flammable Fabrics Act. The first outlawed the material in the cowboy costumes in the early 1950's, and it remains in effect today. Two standards regulate the flammability of carpeting, specifying a test method that simulates a dropped cigarette or cigar and forbidding the sale of a material that sustains combustion beyond a certain radius from the point of ignition.

A fourth standard, published in 1971, governs the flammability of children's sleepwear, a category of textile products frequently involved in fatal fires. The fabrics then most in use, cotton or cotton-polyester blends, did not pass the test. One option open to manufacturers was to treat these fabrics with fire-retardant chemicals. In the early 1970's a number of treatments were on the market or in advanced stages of development. Most of these involved the element phosphorus; some included both phosphorus and either chlorine or bromine. They engaged as working principles the interference by phosphorus in the decomposition reactions of cellulose that produce fuel gases, causing it to form instead slow-burning char, and the disruption by chlorine or bromine of the free-radical reactions in the flame (*see the box on page 144*). All fabrics treated had to retain fire resistance through 50 launderings. One mode of treatment for polyester fabrics required no special equipment or process control; the retardant could be easily applied and heat set. From the retardant's formal name, tris-2,3,-dibromopropyl phosphate, specifying the three brominated propyl groups attached to the phosphorus atom:

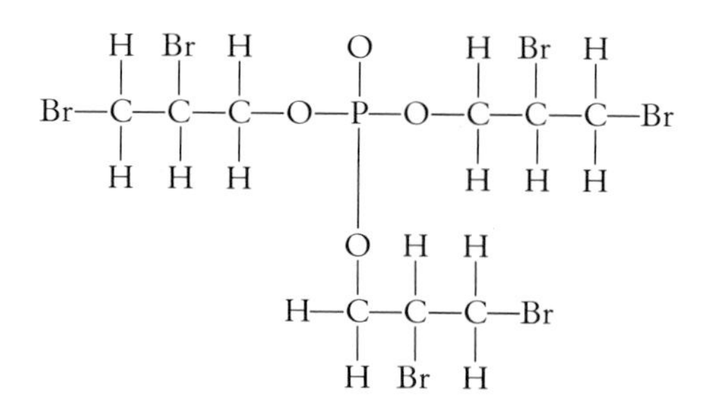

the compound became known to the trade and later in the press as tris.

Tris met the standard and headed for success in the marketplace. Then another Federal laboratory found that this retardant induced cancer in experimental animals. There was panicky argument about whether tris could cause cancer in children by absorption through the skin, how much of it would have to be absorbed, after how many washings, and so on. The Consumer Product Safety Commission was accused of endangering children by its endorsement of tris—an unfounded charge that nevertheless stuck because the desired fire retardance could be achieved by other treatments. After the banning of tris by the Consumer Product Safety Commission, the tainted garments had to be recalled. Litigation related to the episode, in particular to compel the Federal government to indemnify the textile manufacturers, continues to this day.

Despite the ruckus over tris, the standard for flammability of children's sleepwear remains in force. Compliance seems to be general. Burn treatment centers no longer see many victims of fire involving children's sleepwear.

The fifth Federal flammability standard, published in 1972, governs the materials that go into mattresses. From the extensive studies conducted under the Fire Research and Safety Act it had been established that fires originating in mattresses occurred frequently enough to warrant special measures of control. Study of mattress fires showed two things. The first, in confirmation of a common impression, is that cigarettes (and occasionally cigars and burning pipe tobacco) rather than open flames start most of these fires. Second, it is the mattress not the bedding that is the problem; a cigarette will smolder right through blankets and sheets to the mattress without setting the bedding aflame. The mattress standard is straightforward and calls for a real fire test. Mattresses sampled from production are tested by placing lighted cigarettes at random on the surface. If sustained smoldering does not occur, the mattress complies. Compliance is relatively easy and is effected by a careful selection of materials. Polyurethane foam mattresses generally pass the test. Once the smoldering starts, it continues for a long time, producing large amounts of combustion products, which often build to lethal levels. Open flaming may or may not occur. To extinguish a mattress fire, the mattress often must be totally immersed in water.

The analogous problem of the ignition of fires in upholstered furniture by cigarettes has a different history. In 1975 the figures accumulated by the fire researchers were subjected to exhaustive analysis at the Center for Fire Research, which had been established at the National Bureau of Standards. The predominant scenario for the fatal residential fire put the smoker at home at night in an upholstered chair at center stage. Under its mandates from the

A section from a National Bureau of Standards analysis of possible fire scenarios highlights the scenario that causes the largest single number, nearly one-third, of all fire fatalities. This exhaustive study traced the genesis of 5,040 different outcomes.

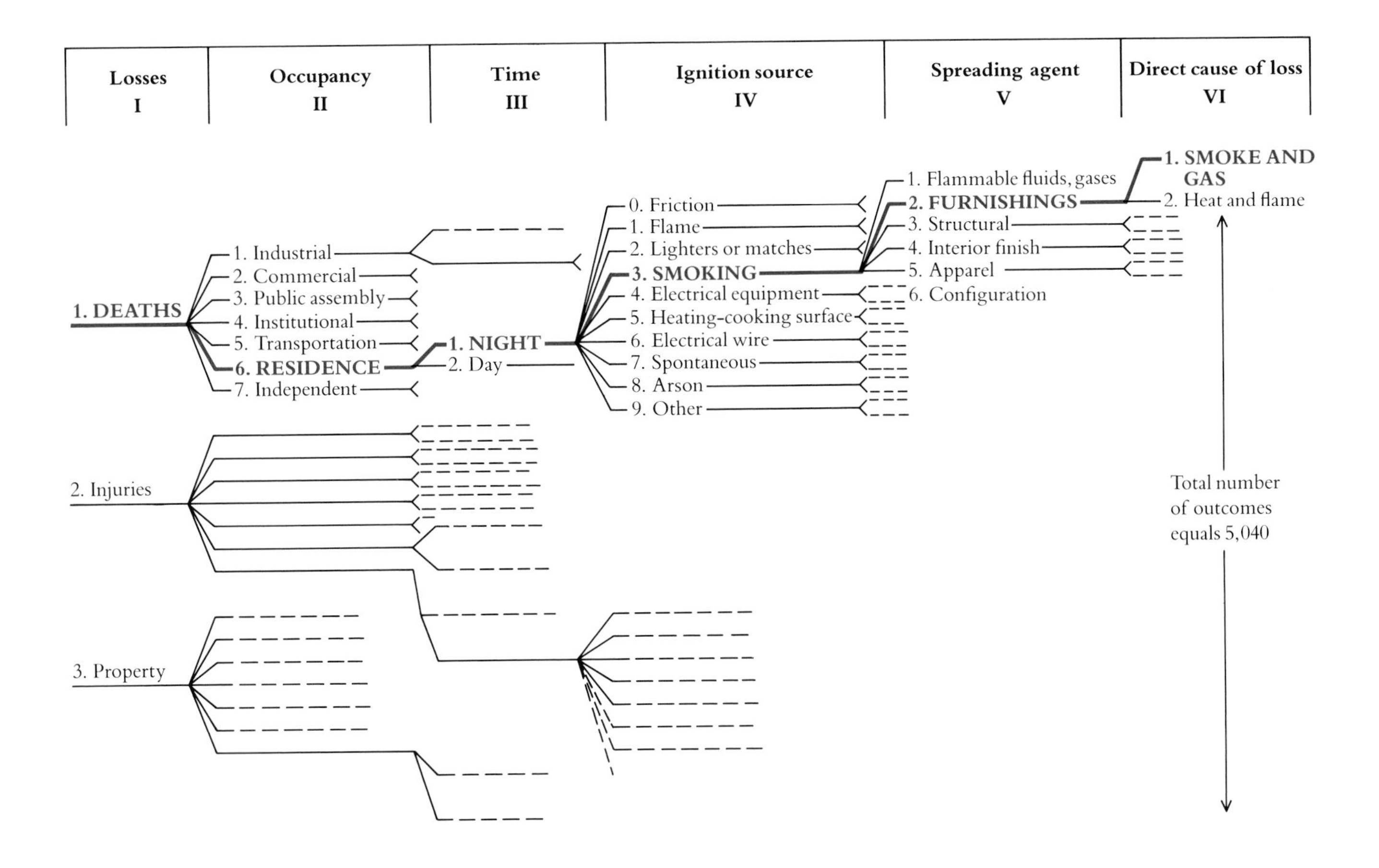

Fire Resistance in Materials

Inherently fire-resistant polymers have been fashioned for use in the clothing of people whose occupations expose them to the hazards of fires (*see the box on page 148*). The more familiar polymers, natural and synthetic, in common use for clothing and house furnishings such as appliance housings and furniture made of molded polymers may be upgraded in fire resistance. The strategy is to add to the polymer—in its synthesis or by impregnation or by coating—certain elements that share the curious property of interfering at one stage or another in the chemistry and physics of combustion. Principal among these are phosphorus, antimony, boron, chlorine, and bromine. The phosphorus compounds act by altering the decomposition of the fuel. For cellulose the mechanism is clear: the phosphorus compound decomposes in the heat of the fire to form phosphoric acid, which then reacts with cellulose to produce large amounts of carbon char, at the expense of the reactions that normally would generate combustible gases. Such treatment makes a material hard to ignite with a small ignition source. The reactive halogens, chlorine and bromine, function in the chemistry of the flame itself as "radical poisons," terminating radical chain reactions that occur in the flame. The compound containing the halogen first vaporizes and then decomposes to intercept radicals essential to the propagation of the flame reactions. An example is the removal of a hydrogen free radical by a bromine compound:

$$RBr + H\cdot \rightarrow HBr + R\cdot$$

In this reaction the sluggish organic fragment, $R\cdot$, replaces the hydrogen radical.

The cellulose textiles, cotton or rayon (from wood pulp), are most often treated with phosphates or borax–boric-acid mixtures. To secure resistance to water in laundering, the phosphorus may be locked into the cellulose by reacting the cellulose with a phosphorus-containing compound or, in the synthesis of rayon, by polymerizing the monomer with a phosphorus-containing monomer. This technology is employed in making textiles for children's sleepwear, which is almost the only protective measure established by the upsurge of national concern about fire at the beginning of the last decade, as recounted in this chapter.

Lumber for inside use may be made fire resistant by pressure treatment with solutions of ammonium phosphates and borax.

The fire-retarding virtues of chlorine and bromine can be incorporated in the original synthesis of the synthetic fiber or film. Thus chlorine may be incorporated by reacting it with polyethylene:

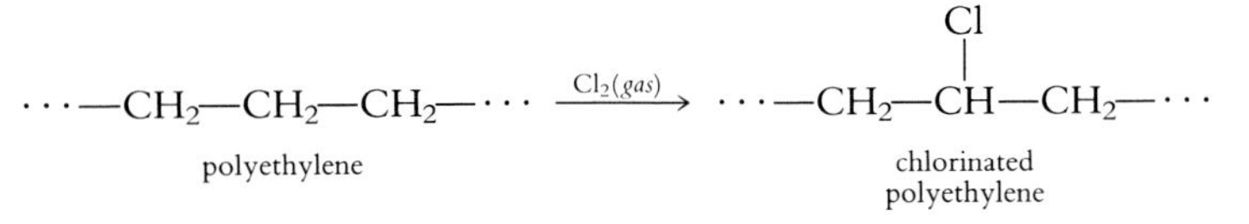

(The ratio of chlorines to carbons in polyethylene is variable.) It is difficult, however, to incorporate as many as one chlorine for each pair of carbon atoms. One way to accomplish that is to start with chlorine in the monomer:

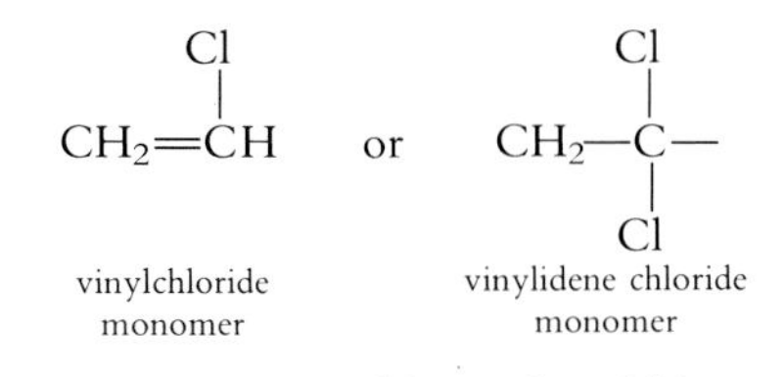

Both of these materials resist ignition and burn sluggishly and are chemically inert as well. These virtues have brought wide use of polyvinyl chloride in plumbing pipes. Chlorine polymers are not very heat resistant, however, decomposing at relatively low temperatures. Their fragments do not sustain a flame in ordinary air (*see the table below*) in the absence of external heat.

Chemical Stability of Some Polymers to Heat

	Percent Cl	Decomposition temperature range (°C)	Self-ignition temperature (°C)	Percent oxygen level to support combustion
Polyethylene	0	335–450	349	17.4
Polyvinyl chloride	57	200–300	454	45–49
Polyvinylidene chloride	73	225–275	532	60

Synthetic polymers that contain oxygen in their backbone, such as the polyesters and polyurethanes, are often fire retarded with phosphorus. Polyurethane foams for use in insulation are usually treated this way.

At some level of imposed heat, however, all carbon compounds will burn. Resistance to ignition and burning under one set of conditions does not mean resistance under all conditions. Most common polymers treated with fire retardants are, indeed, harder to ignite than untreated polymers and require some increase in external heating to keep their combustion going. This difference is not meaningful in all applications.

Flammable Fabrics Act and the Fire Prevention and Control Act, the Consumer Products Safety Commission had to find a way to intervene. Congress had earlier removed cigarettes from the commission's assigned responsibilities. This left the ease of ignition of furniture as the target.

A test and a standard covering the ignition of sample upholstery constructions were developed by the Center for Fire Research. These were immediately challenged by manufacturers of upholstered furniture. The test requires resistance to ignition by lighted cigarettes and was a real fire test in that sense. But since nearly every piece of upholstered furniture is unique (unlike mattresses), the government proposed the use of small mockups of the fabric, interlayer, and padding composite and a scheme for rating fabrics separately to reduce the cost of testing. The industry argued that the cost of compliance would be excessive and that certain fabrics would be removed from the market, notably plush and brushed cottons on the expensive end of the line, thereby reducing consumer choice. The government argued that cost increases would be only a few dollars an item, that many combinations of fabric and padding then in use would pass the test, and that the benefits to society outweighed the costs. Industry spokespeople argued that smoke detectors offered a cheaper alternative. The commission put standard making on hold.

With the Federal standard in limbo, the industry developed a voluntary action plan using less complex and less stringent tests and standards. The industry action group has not yet secured compliance by all manufacturers. Recent (1981) results of Consumer Product Safety Commission testing showed progress but still many noncomplying products being sold. In the course of this long-drawn-out confrontation, the most extensive cost-benefit analysis ever performed in the fire-safety arena was published by the National Bureau of Standards and SRI International. In this study the prospective effect of the standard and of smoke detectors on the loss of life was considered separately and jointly at various efficiency levels and offset by compliance costs. The results seemed to favor the standard by a modest margin.

The failure to secure a test and standard for the flammability of upholstered furniture apparently signifies that no new Federal mandatory fire-protection standards will be published under the Flammable Fabrics Act or under any related Federal statutes. It may represent the end of major standard making by the Consumer Product Safety Commission. Yet the case for doing something dramatic about the cigarette–upholstered-furniture combination was, and is, undoubtedly the strongest and best documented of all.

History shows that the tide of consumerism that produced the fire legislation and the Consumer Product Safety Act crested in the early 1970's and has receded. Apparently, both the Consumer Product Safety Commission and the U.S. Fire Administration came into being somewhat after the crest. The U.S. Fire Administration never came close to receiving the funding conceived for it in the legislation. Although the Consumer Product Safety Commission was funded sufficiently for a while, and did succeed in establishing some significant fire-preventing standards, it must have undertaken the initiative with respect to upholstered furniture too late: during the ebbing of consumerism in Washington.

The story serves to dramatize the assertion that, for a long time to come, it will be difficult to enlist public authority in developing and enforcing measures to reduce fire losses in one- and two-family residences. With "deregulation" the watchword, the Federal government is not going to impose flammability standards on the products that are so often the first items ignited in fatal fires. The climate of opinion now says, apparently, that the people do not want their government to protect them from themselves or from heedless manufacturers.

At the first branching point in the fire-scenario decision tree, therefore, we are compelled to turn away from the "nonignition" option and to contemplate the possibility of "ignition." Suppose a fire does get started in your house!

Since fire fatalities and injuries result largely from the products of combustion, it follows that early detection of such products should reduce these human losses. The challenge is to detect the fire and sound the alarm soon enough to provide ample time for escape. For an alert, fully functional adult, only a few seconds may be needed. For a handicapped person, for one who is asleep, or for the very young or very old, several minutes may be necessary. An early warning that provided, say, at least five minutes safe time would be desirable.

This safe time is for the protection of occupants. If protection of the house is the object, then the warning time must exceed the response time of the fire department. This response varies greatly from city to rural areas, from as low as one to six minutes in the city to 10 or more in the country.

Since smoke and gases per se are not nearly as damaging to property as heat and flames, the heat detector was developed years ago for use in protecting commercial and industrial buildings. Historically the focus in fire prevention has been on stopping large fires. To do this, efforts must be concentrated on detecting and suppressing fire as opposed to sensing toxic gases. Two types of heat detector are used: fixed temperature and rate of rise (of temperature). The fixed-temperature detector operates much as a sprinkler head. It has an element that at a given temperature closes or opens a control circuit. The simplest type is a fusible link, an element that melts at a given temperature. The alarm circuit is activated by a spring or another element. Other types include frangible glass bulbs and bimetal units using two joined metals of differing thermal expansion, much as in simple bimetal thermometers (*see the figure on page 150*). These detectors set off an alarm no matter how slowly the temperature changes. Moderately high temperature can arise from benign causes and produce false alarms. In most fire situations the temperature rises

Fire-resistant Clothing

The burning of clothing, as so many cruel accidents to small children have demonstrated, serves as a way to bring fire into intimate contact with flesh. Against that hazard in ordinary domestic existence the treatment of the commonly used textile fibers with fire retardants suffices. In fact, such treatment of children's sleepwear has substantially reduced the annual fire-casualty rate among youngsters.

Fire is a hazard that attends certain occupations, however, and the exposure tends to be severe. People employed in the liquid-fuel industries, from the field to the refinery and on the decks of tankers, work in the constant danger of flash fire. Racing-car drivers and airport ground crews share such danger. Exposure to fire is, of course, the profession of fire fighters.

A heat-instrumented manikin was photographed after four seconds of exposure to a propane flame (far left), dressed in coveralls made of aramid (middle left), cotton (middle right), and cotton-polyester fiber (far right). The aramid coveralls remained almost undamaged from the envelopment in flame; the cotton and cotton-polyester coveralls continued to burn.

For the protection of such people textiles are now available made of inherently fire-resistant synthetic-polymer fibers. The materials are expensive, and they do not make up into attractive fabrics for everyday wear. They serve well, however, in coveralls, flight suits, and uniforms, and they have a record of saving their wearers from calamity.

One family of fibers, marketed as the aramids by the Du Pont Company, consists of aromatic (benzene-ring-containing) versions of nylon:

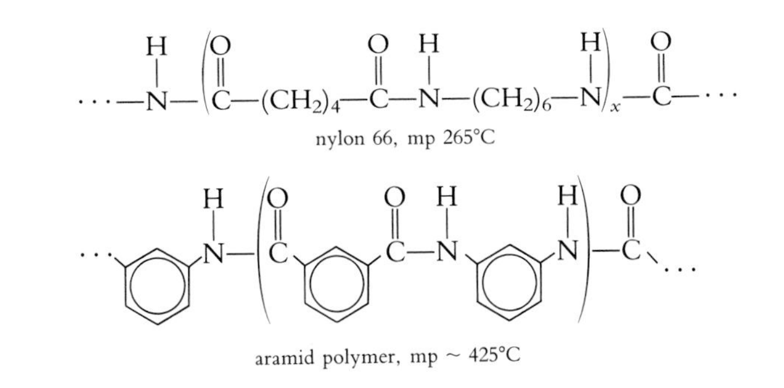

nylon 66, mp 265°C

aramid polymer, mp ~ 425°C

Nylon melts easily (and can cause severe burns by so doing, without burning) and burns with the help of sufficient heat from a fire. The aramid fiber does not melt or burn, but chars and stiffens.

Whereas the aramid fiber has a hydrogen on its nitrogens, offering oxygen a site of oxidative attack, another structure—the aromatic imide polymer—does not:

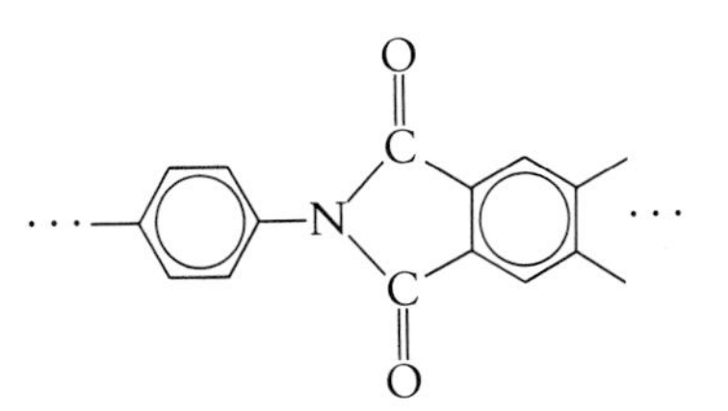

Exposed to direct flaming, it shrinks and blackens but is not consumed and does not produce much smoke.

For fire tests of such textiles, the U.S. Air Force has developed a six-foot one-inch manikin known as Thermo-man, with 122 surface heat-flux sensors distributed over its fiberglass-plastic body. In one test the manikin, dressed in cotton and then in aramid coveralls, was exposed to a four-second propane flash fire. In cotton the instruments indicated second- and third-degree burns over 99 percent of the body surface; in cotton-polyester, 75 percent; and in the aramid garment 33 percent, with very much reduced third-degree burns.

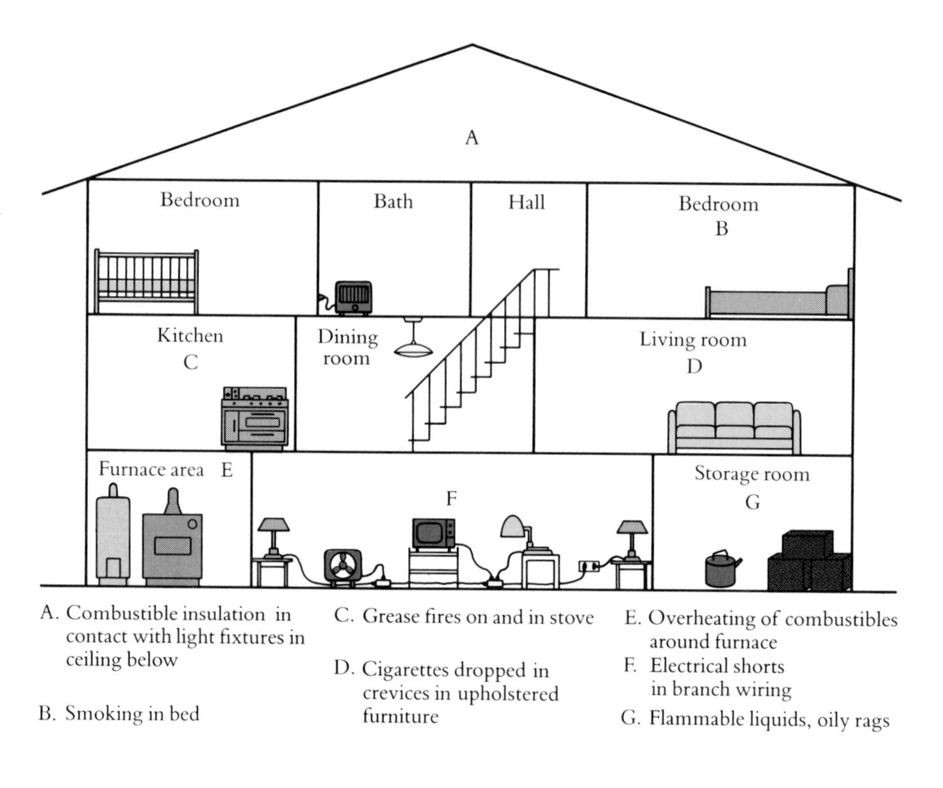

Fire hazards in the household are highlighted in this schematic diagram. Starting in the cellar, while a furnace in good repair is not a notable hazard, the storage of combustibles near it exposes them to possible overheating. In the storage room, flammable liquids and oily rags are surrounded by less combustible but fuel-rich materials. The stove in the kitchen can be the site of grease fires in the oven and on the burners. Combustible insulation in contact with wiring, such as that of a ceiling fixture, can transform a short circuit into a fire. Upholstered furniture, smoldering from a dropped cigarette, figures in the most common fatal fire scenario. The next most frequent site of ignition is the bed, again set afire by a cigarette. Electrical heaters may draw too much current and overheat household wiring; the wiring may be similarly overloaded by too many extension cords from one outlet.

rapidly; for these fires a detector based on the rate of temperature rise is a more discriminating device. Most of the rate-of-rise detectors contain an air chamber in which the pressure developed by heat serves to close a circuit to the alarm. A small vent is provided, sized so that on slow heating, the gas, expanding in the chamber, escapes without closing the circuit, whereas on rapid heating, the gas cannot leave fast enough to prevent pressure development and alarm.

Heat detectors have simple circuits, are reliable, are relatively free of false alarms, and have a long and distinguished record of property protection. By their very nature they are not sensitive to smoke particles and gases at moderate temperatures, as might be the case with a fire that breaks out at a distance from the detector or with a smoldering fire. Detection of low-temperature smoke and gas requires another approach, one taken only relatively recently.

From the frontiers of solid-state physics and nuclear physics have come two new instruments for smoke detection. One relies on sensing the scattering of light by smoke particles and is known as a photoelectric detector. The

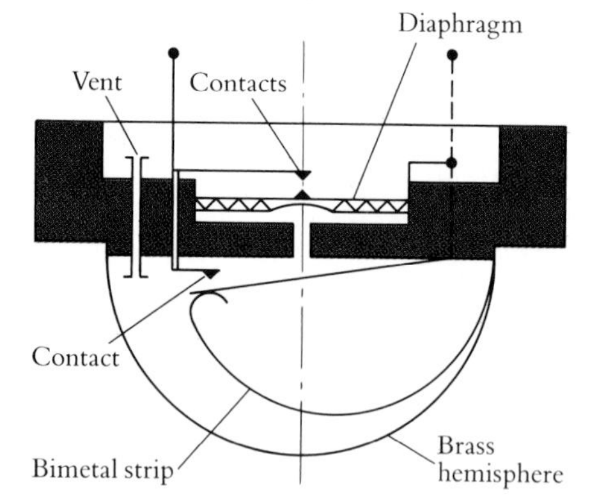

A heat sensor employs a "bimetallic" element, a laminate of metals with different coefficients of expansion that bends on heating. At the threshold temperature its bending closes the alarm circuit. The brass chambers and vent make it sensitive only to rapid changes in air temperature.

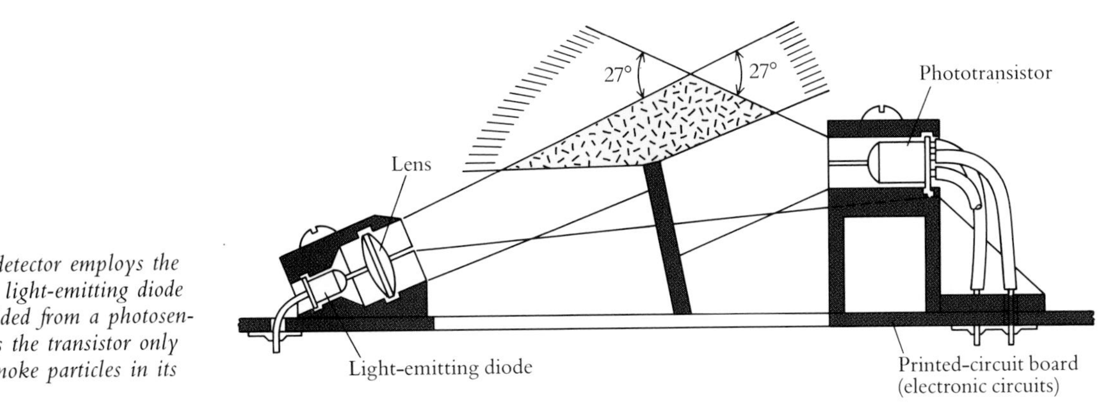

A photoelectric smoke detector employs the low-current-consuming, light-emitting diode at left. The light, shielded from a photosensitive transistor, reaches the transistor only when it reflects from smoke particles in its path.

second relies on the ionization of gas molecules and particulates to produce changes in electric conduction in a test chamber and is termed an ion-chamber detector. Each type has its advantages and can be adjusted to be very sensitive.

The photoelectric detector makes use of the well-known phenomenon that small particles will scatter light. Known as the Tyndall effect, light scattering is commonly noted when the path of a sunbeam is made visible by light reflected from dust in the air. In its simplest form, the photoelectric detector contains a light-emitting diode, a smoke chamber, and a light detector. Light strikes the smoke particles and is scattered in all directions but not equally. For smoke particles and the wavelengths of light used in these detectors, scattering in the forward direction is the most sensitive measure of particle concentration (*see the figure above*). Because the diode operates in the red portion of the spectrum and its intensity is low, the receiver or detector of scattered radiation has to be specially designed. Photovoltaic cells based on silicon semiconductors are now in use in combination with low-cost integrated electronic circuits to amplify the signal and process it for the alarm circuit.

The second variety of smoke detector in wide use is the ion-chamber detector. These devices have a small source of ionizing radiation within a chamber containing two plates with an applied electric potential (*see the figure on page 152*). Air molecules or smoke particles enter the chamber, are struck by the radiation, and are ionized; that is, electrons are removed from some of the molecules or particles and are attached to others, producing positively and negatively charged molecules or particles in the chamber. These flow to the negative and positive plates respectively of the chamber, thereby setting up an electric current. The system is designed so that if only normal air molecules are present, no alarm is sounded. If larger gas molecules or liquid or solid particles flow into the chamber, they may acquire a charge, but because of their larger inertia (larger mass), compared with that of air molecules, they may flow out

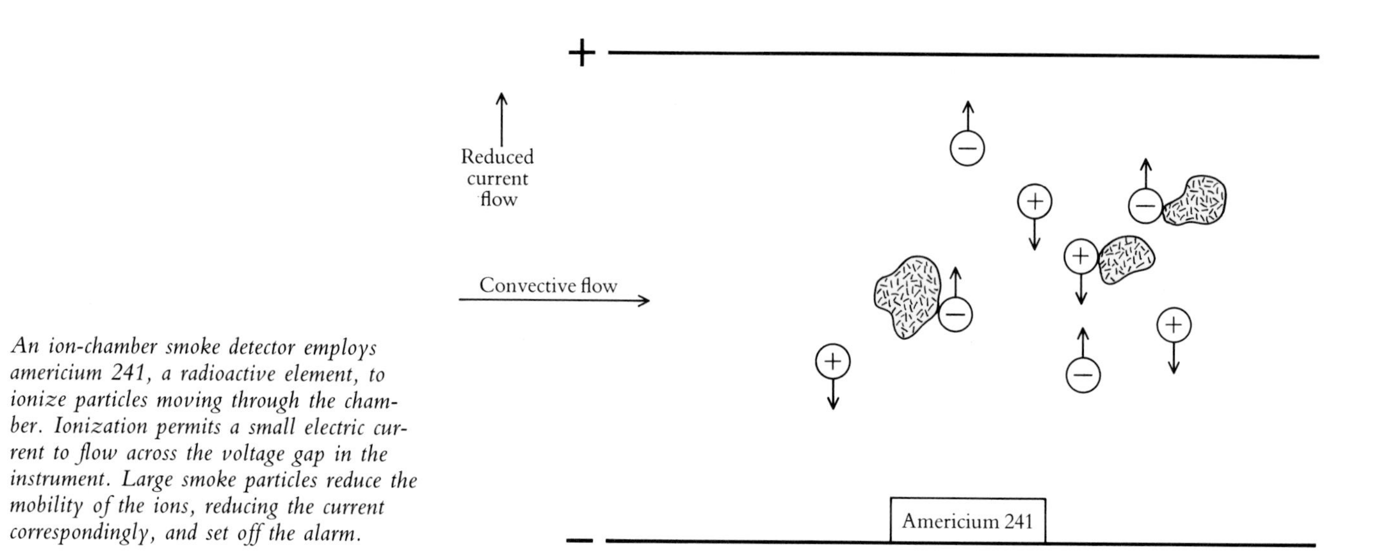

An ion-chamber smoke detector employs americium 241, a radioactive element, to ionize particles moving through the chamber. Ionization permits a small electric current to flow across the voltage gap in the instrument. Large smoke particles reduce the mobility of the ions, reducing the current correspondingly, and set off the alarm.

of the chamber before they reach the charged plates. The effect is to reduce the current flow, and the circuit is designed so that the alarm is sounded when a given reduction in current occurs.

The change in current is directly proportional to the number of particles multiplied by their diameter, so that the more large particles present, the greater the change in current. It is evident that very high velocities of air through the chamber could cause an alarm by blowing even the small charged air molecules out before they reach the plates, and that very low velocities would permit the large smoke particles to reach the plates rather than to move on out of the chamber. The detectors are designed so as to make trade-offs between these effects.

Studies of the response of smoke detectors to test fires show that the ion-chamber detector is more sensitive to gaseous combustion products and the photoelectric detector is more sensitive to particulates. Thus for clean-burning fires with little visible smoke, the ion-chamber device is more effective; for smoldering fires producing large quantities of visible particles, the photoelectric detector responds somewhat better. However, these differences are easily discerned only in experimental situations in which the detectors are placed fairly close to the fire. When the detectors are placed farther from the source, the differences between the two types become small. The National Bureau of Standards has concluded that there are not sufficient differences between the two types to warrant selecting one over the other for most applications in the home. Neither type is suitable in the kitchen because nonhazardous smoke is common there. If an alarm system is desired for the kitchen, a heat detector should be used. There are other special cases in the home where one or another type of detector is preferred because of the circumstances most likely to be encountered.

The ion-chamber detector contains a source of ionizing radiation, most commonly the radioisotope americium 241. In the detector as it is normally constructed, the radiation source is well protected so that exposure of people to it is prevented. However, if the detector is damaged in a fire or in any other way, the radiation source might be exposed. The question has been raised as to whether there is an increase in risk by using such radiation sources, particularly when the photoelectric devices appear to be effective. A longer-term question is what effect will there be when, at some time many decades from now, these detectors are discarded into our junkyards and landfills or burned in incinerators. For the moment, at least, the risk is held to be not sufficient to warrant restrictive rule making regarding these ion-chamber devices.

In contemplation of the possibility of a fire in the house, it is best if one has already given thought to the combustibility of materials in the ceiling, wall, and floor coverings. Although such materials are regulated for buildings with public access by some local codes, they go unregulated for use in households. Carpeting is the one material that has come under control by Federal test and standard. It has been regulated as to ease of ignition, but not with respect to the contribution it may make to a hot, spreading fire started elsewhere. Wall and ceiling coverings are simply not controlled. You can have as much highly waxed wood paneling as you like. And we have just seen that the combustibility of furniture is not likely to be further regulated. So we should not look to this part of the fire-protection strategy for much help with the home fire.

Another component of the strategy is to impede fire progress by building design. In large buildings, we have seen that enclosing stairways is a major fire-containment technique, as is also providing compartmentation on a given floor. In large, sensitive structures, fire-rated barriers are used to isolate sectors. These include both wall and door designs for half-hour fire resistance and more. Of course, the structural elements in large buildings are all fire rated.

In the one- and two-family dwelling, one cannot reasonably expect ever to see fire-rated assemblies. They are expensive. What we have seen in the past and could see again is both enclosed stairways and compartmentation within floors. In old houses built in the days before central heating, stairways had doors at the bottom or were otherwise shut off to keep heat on the floor where it was wanted. Likewise, each room had a door that could be shut.

I live in a 200-year-old farmhouse. The original stairway is narrow and steep and has a door flush in the wall entering the front room. It is also closed at the top where it enters a bedroom. The "new" front stairs are open. Similarly, every room is provided with a door frame and each originally had a door

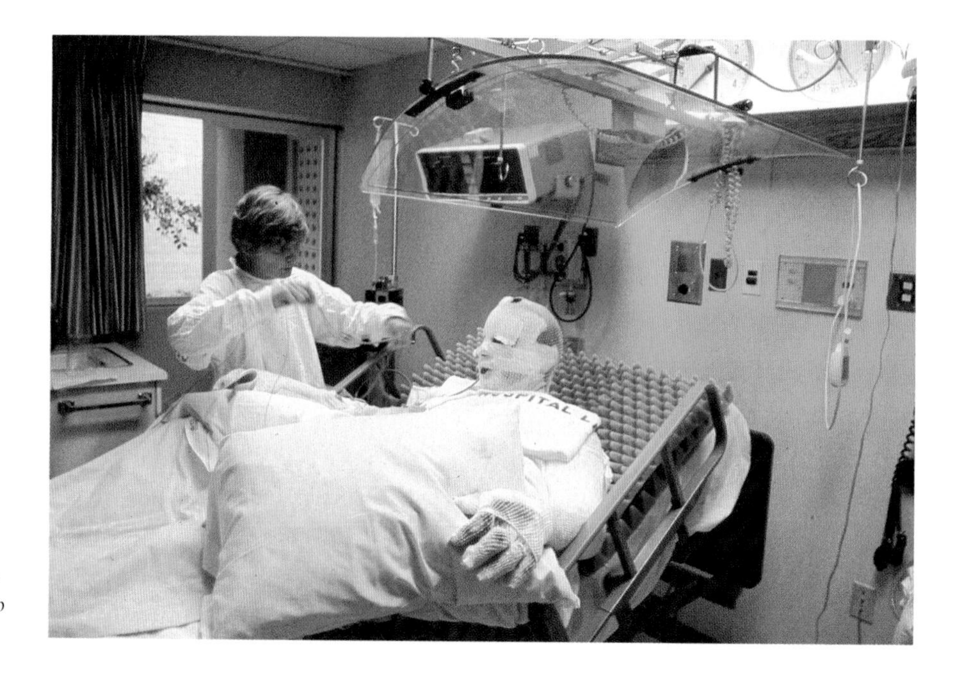

Burns are the most severely taxing and painful injury the body can sustain and are medically and surgically the most difficult to manage.

installed, as evidenced by the presence of hinge recesses and latch receptacles. All the doors have been removed downstairs. Since the price of home heating oil went up by a factor of about 5 in the past decade, I have often wished I had the doors in place and individual room heating. Unfortunately the house has three-level, hot-water zone heating. (I have, however, installed a barrier on the "new" stairs to hold the warm air in the principal living space.)

I believe that the old-fashioned floor plan with rooms closed off was a safer environment with respect to fire than today's wide-open floor plans with no barriers. These designs are matters of taste as well as issues for heating and safety. How much additional time unrated doors would provide is not clear, but even a minute or two can make the difference. Of course, the doors would have to be kept closed, especially at night, and one would have to know, by hearing a previously installed and reliable alarm, that a fire was growing behind one of those doors.

Further on in the decision tree, we come to the suppression of the fire. Ideally, it would be detected and suppressed at the same moment by a sprinkler system. The residential sprinkler system has been the dream of many fire-protection specialists. Sprinklers as they are now installed in commercial, industrial, and large public occupancies are reliable but very expensive. Reliability is over 99 percent for a properly installed and maintained system. The cost for a single-family dwelling of a commercial-grade system would be on the order of a 2 percent addition to the building costs ($2,000 or so). Such an increase in cost appears not to be acceptable. Studies are being made, therefore, to see if a lower-cost system with necessarily reduced reliability makes any sense. The principal problems are how to obtain good knockdown performance with small water lines, relatively low pressure, and lightweight piping and hangers. The sprinkler heads do not pose a problem. Considerable progress on this has already been made and home sprinklers are now at least feasible. The National Fire Protection Association is writing standards for residential sprinklers that will help homeowners in selecting systems. There is almost no chance, however, that sprinklers will be broadly required by code for one- and two-family homes.

The last step in fire-protection strategy brings us to the occupants and their rescue. Their safety was, of course, the first concern; with the sounding of the alarm, they were to begin their escape. For less mobile members of the family, handicapped, infant, or very old members, escape may be an arduous or impossible undertaking. For these cases, immediate access to the outside is

A sprinkler head affords a fail-safe active response to the heat of a fire. A link metal with a low melting point (at right just above the round horizontal deflecting plate) holds two levels of a valve assembly in place, confining water under high pressure in the pipe. When the metal yields to heat, the valve opens; water impinging on the plate sprays into the surrounding area.

one answer: a ground-floor room with a doorway to the yard. An expensive alternative would be to provide fire-rated protection all around an upper story room and, if heating is by hot air, an independent heating and cooling system. This option is the equivalent of the protect-in-place approach required in hospitals and other health-care facilities. For existing dwellings, the protect-in-place concept is not feasible and would be very expensive. This is also true even when building a new home. My own view is that investment in a home sprinkler system makes more sense.

The means of escape from most two-story houses are windows, a single central stairway, and two doors to the outside from the ground floor. The stairway is neither separated from the rest of the house nor enclosed. Often it is a principal conduit of the fire and hot combustion products and is not available for escape.

Standard fire-safety advice is to sleep with the doors to the bedrooms closed, have loud smoke-detector alarms just outside the doors, feel the door before opening it, and if it is hot, go out the window. In the press of a fire, able-bodied adults will probably go out fast. Very young, very old, and infirm people cannot. For these, their only choice is to keep the door tightly shut, go to the window if they are able, and make their presence known. They must count on other members of the family to do the rest.

Every family should have an emergency plan: each member should know where to exit and where to meet outside. In our family's case we have a large, outdoor all-night light, equal to a streetlight, to illuminate the parking area and barnyard. We shall meet under that light in the event of a disaster.

There are rope ladders on the market that one can install under a windowsill and throw out in the event of a fire. For a bedroom with no porch roof below the windows, such a ladder would be a good investment. For anyone living in a third-floor garret, a rope ladder is imperative.

The principal measure presently being taken toward the suppression of the residential fire in the United States is public education. The National Fire Protection Association and the U.S. Fire Administration conduct significant programs. The NFPA has devoted substantial resources to competitions and to campaigns on television. Most of this effort has been aimed at telling children what to do in the event of a fire. The effort appears to be useful and is producing documented savings of life. Care and handling of flammable liquids is an easy message to get across. The ease of ignition of common furnishings and the speed and development of a fire are less easily taught; certainly no one expects to convert the population into fire-protection engineers en masse. However, there is clearly room for more detailed and specific information for the general public and probably for more outright fire-safety propaganda. There are currently no agencies, public or private, with enough money to mount a sustained television propaganda effort of the magnitude raised a few years ago on smoking and cancer.

A mobile-home community in the Los Angeles area creates a new, more densely crowded pattern of suburban U.S. single-family dwelling. Supplying presently about 25 percent of the country's new housing, these prefabricated houses have become progressively less mobile as they have increased in floor space and acquired extensions for carports and sheltered outdoor living spaces. Because they are constructed under strict Federal regulation of potential fire hazards, these homes are rated safer than the conventional detached dwelling. Community developers provide such common essential utilities as electric power, water, and sanitation, plus such amenities as swimming pools and community centers.

Closer to the realm of action, there is no governmental effort at any level to search out and systematically eliminate fire hazards in the products sold for, or installed in, the home. Evidently 5,000 or so deaths in fire is an acceptable loss for us to bear in the United States. It is only a fraction of the loss to automobile accidents. Were it not for the better record of fire safety in other countries, a record recited at the outset of this book, there the matter could rest. But knowing of our poor record, and of our ability to solve problems once we decide to do so, many believe the matter should not rest where it is. Perhaps there will be a resurrection of public interest, in another decade or after a particularly catastrophic fire.

For Further Reading

Chapter 1

Fire Journal, National Fire Protection Association, Batterymarch Park, Quincy, Mass. 02269. Published bimonthly, this archival journal carries each year in its September issue a full summary of fire-loss statistics for the United States.

America Burning, Report of the National Commission on Fire Prevention and Control, U.S. Government Printing Office, 1973. 177 pp. This report led to the passage by Congress in 1974 of the Federal Fire Prevention and Control Act, which set up the U.S. Fire Administration, the Fire Academy for training fire fighters, and an expanded program of fire research at the Fire Research Center at the National Bureau of Standards.

Chapter 2

M. Faraday, *The Chemical History of a Candle,* Thomas Y. Crowell, New York, 1957. 158 pp. Classic lectures by Faraday at the Royal Institution in London to an audience of 12-year-old boys.

Harvard Case Histories in Experimental Science, Vol. 1, J.B. Conant, ed., Harvard University Press, Cambridge, Mass., 1957. 321 pp. This volume contains detailed information on three topics of interest: Robert Boyle's experiments with gases, Benjamin Thompson's disposition of the caloric theory, and the discovery of oxygen and Lavoisier's formulation of modern chemistry. Fascinating reading, aimed at the nonscientist.

References in physics and chemistry are legion. For a start, consult modern high school textbooks. For more specialized studies, I recommend the following:

F.W. Sears, M.W. Zemansky, and H.D. Young, *College Physics,* 4th ed., Addison-Wesley, Reading, Mass., 1977. 751 pp. A good general text.

The Berkeley Physics Course, five volumes. McGraw-Hill, New York, 1967–1973. Detailed two-year curriculum. A good way to learn physics, but not for the casual reader.

R.P. Feynman, et al., *The Feynman Lectures on Physics,* Vols. I–III, Addison-Wesley, Reading, Mass., 1963–1965. A two-year curriculum given at the California Institute of Technology, presented the way a physicist thinks and with personal insights that are often gems. Tough going but rewarding.

I. Glassman, *Combustion,* Academic, New York, 1977. 275 pp. Lots of chemistry, some fluid mechanics and transport phenomena, and not too much mathematics.

A.G. Gaydon and H.G. Wolfhard, *Flames,* Halstead Press, Wiley, 1979. 401 pp. Much data on flames, especially on experimental techniques for studying them. Not for beginners.

R.M. Fristrom and A.S. Westenberg, *Flame Structure,* McGraw-Hill, 1965. 424 pp. A good discussion of flame chemistry, with enough engineering to round it out. Also not for beginners.

J.W. Lyons, *The Chemistry and Uses of Fire Retardants,* Wiley (Interscience), New York, 1970. 462 pp. A review of how fire retardants work and common ways to treat materials ranging from paper and wood to advanced synthetic polymers. Not too advanced.

Chapter 3

D. Bradley, *Count Rumford,* Van Nostrand, Princeton, N.J., 1967. 176 pp. A nontechnical short life of Benjamin Thompson. Good for a quick read.

B. Franklin, *The Autobiography of Benjamin Franklin,* Signet Books, New American Library, New York, 1961. 350 pp. Franklin never finished this; it stops before the American Revolution. It covers the Pennsylvanian fireplace and the founding of the Union Fire Company and many other public bodies in Pennsylvania that persist to this day. A wonderful story.

D. Macaulay, *City: A Story of Roman Planning and Construction,* Houghton Mifflin, Boston, 1974. 112 pp. An architect's rendering of the Roman baths, among others. Large libraries might have one of several translations of Vitruvius' review of Roman building practices.

Chapter 4

L. Cummins, *Internal Fire,* Carnot Press, Lake Oswego, Wis., 1976. 351 pp. An entertaining history of internal combustion engines to about 1900. For the general reader.

Chapter 5

Fire Research on Cellular Plastics: The Final Report of the Products Research Committee. 213 pp. Published in 1980 by the Committee, a nine-member trust set up to implement part of a settlement between the makers of cellular plastics and components thereof and the Federal Trade Commission. A good review of the state of the art of fire research as of 1980. Available through the Center for Fire Research, National Bureau of Standards, Gaithersburg, Md. 20899.

Chapters 6 *and* 7

P. Lyons, *Fire in America,* National Fire Protection Association, Batterymarch Park, Quincy, Mass. 02269, 1976. 244 pp. Lots of photographs and case histories of severe American fires. For the general reader.

General References

Dictionary of Scientific Biography, 16 volumes, C.C. Gillespie, ed., Scribner, New York, 1970–1978. Quick way to enter the literature on famous scientists over the ages.

Kirk-Othmer Encyclopedia of Chemical Technology, 24 volumes, 3d ed., M. Grayson, ed., Wiley (Interscience), New York, 1978–1984. An old standby for me. Reasonably short and understandable write-ups on topics even vaguely related to chemistry and chemical manufacture. Includes items such as boilers and turbines. Can be found in large libraries and most technical libraries.

Handbook of Chemistry and Physics, Chemical Rubber Co., Cleveland, Ohio 44128. Issued in a new edition almost every year since 1913 under a series of editors, this massive compilation of data is a must on the shelf of every practicing chemist or physicist. Because of the large number of users, the price is very low. Recommended for the interested reader who wants to dabble a little and needs an occasional number for a calculation.

Sources of the Illustrations

Drawings by Gabor Kiss

Chapter 1

Chapter opening
Kees Van Dan Berg/Photo Researchers, Inc.

page 2
Kaku Kurita/Gamma-Liaison

page 5
Courtesy of Dr. Thomas Croft

Chapter 2

Chapter opening
"The Amateur Scientist," by Jearl Walker. Copyright © 1978, Scientific American, Inc.

page 11
The Granger Collection

page 14
The Granger Collection

page 19
The Granger Collection (top); Bettmann Archive (bottom)

page 24
Bausch and Lomb

page 32
Copyright © 1982, Scientific American, Inc.

Chapter 3

Chapter opening
Fred McConnaughey/Photo Researchers, Inc.

page 42
Bettmann Archive

pages 44 and 45
Courtesy of Dartmouth College Library, Special Collections

page 48
The Granger Collection (bottom)

page 49
New York Public Library

pages 50 and 51
Courtesy of Dartmouth College Library, Special Collections

page 53
Culver Pictures, Inc.

Chapter 4

Chapter opening
Courtesy of the American Iron and Steel Institute

page 64
John Moss/Photo Researchers, Inc.

page 68
The Granger Collection

page 69
Bettmann Archive

page 71
The Granger Collection

page 73
Courtesy of BBC Brown Boveri, Inc.

page 78
Courtesy of Sandia Laboratories

page 80
Courtesy of Sandia Laboratories

page 82
Courtesy of NASA

Chapter 5

Chapter opening
Courtesy of Dr. Howard Emmons

page 86
Center for Fire Research, National Bureau of Standards

page 94
Center for Fire Research, National Bureau of Standards

page 95
Bill Conduit

pages 100 and 101
Center for Fire Research, National Bureau of Standards

page 102
Courtesy of Factory Mutual Research Corporation

Chapter 6

Chapter opener
A.E. Willey, *Fire Journal,* Vol. 66, No. 4, 1972, page 6

page 112
Bettmann Archive

page 113
Bettmann Archive

page 114
Bettmann Archive

page 116
The Granger Collection

page 117
L. Burr/The Darkroom, Inc.–Liaison

page 118
Courtesy of F.A. Albini, Northern Forest Fire Laboratory, U.S. Forest Service

page 119
Photo from European

page 120
"The Climatic Effects of Nuclear War," by Richard P. Turco, Owen B. Toon, Thomas P. Ackerman, James B. Pollack, and Carl Sagan. Copyright © 1984, Scientific American, Inc.

page 121
Scientific American, October 23, 1852

page 122
Associated Press

page 123
Fred Anderson/Photo Researchers, Inc.

page 124
Associated Press

page 127
Hank Morgen/Photo Researchers, Inc.

Chapter 7

Chapter opener
Georg Gerster/Photo Researchers, Inc.

pages 140 and 141
Center for Fire Research, National Bureau of Standards

pages 148 and 149
Courtesy of Du Pont

page 154
J.P. Laffont/Sygma

page 155
David Sailors

page 157
Georg Gerster/Photo Researchers, Inc.

Index

Other Books in the Scientific American Library Series

POWERS OF TEN
by Philip and Phylis Morrison and the Office of Charles and Ray Eames

HUMAN DIVERSITY
by Richard Lewontin

THE DISCOVERY OF SUBATOMIC PARTICLES
by Steven Weinberg

THE SCIENCE OF MUSICAL SOUND
by John R. Pierce

FOSSILS AND THE HISTORY OF LIFE
by George Gaylord Simpson

THE SOLAR SYSTEM
by Roman Smoluchowski

ON SIZE AND LIFE
by Thomas A. McMahon and John Tyler Bonner

PERCEPTION
by Irvin Rock

CONSTRUCTING THE UNIVERSE
by David Layzer

THE SECOND LAW
by P. W. Atkins

THE LIVING CELL, VOLUMES I AND II
by Christian De Duve

MATHEMATICS AND OPTIMAL FORM
by Stefan Hildebrandt and Anthony Tromba